METALLURGY

First paperback printing 2010

Library of Congress Catalog Number: 2010003969
ISBN: 978-0-202-36361-5
Printed in the United States of America

Library of Congress Cataloging-in-Publication Data

Dennis, W. H. (William Herbert), b. 1904.
    [Hundred years of metallurgy]
    Metallurgy : 1863-1963 / W.H. Dennis.
        p. cm.
    Originally published: A hundred years of metallurgy. Chicago : Transaction
    Publishers, c1963.
    Includes bibliographical references and index.
    ISBN 978-0-202-36361-5 (alk. paper)
        1. Metallurgy--History. I. Title. II. Title: 100 years of metallurgy.

TN615.D4 2010
669--dc22

                                                                    2010003969

METALLURGY

1863-1963

W.H. Dennis

AldineTransaction
A Division of Transaction Publishers
New Brunswick (U.S.A.) and London (U.K.)

CONTENTS

Preface   ix

1. Introduction   1

Alloy Steels — Iron — Non-Ferrous Metallurgy — Ore Dressing — Pyrometallurgy — Electrometallurgy — Electric Smelting
— High-Frequency Induction Melting — Hydrometallurgy — Shaping of Metals — Forging — Rolling — Extrusion — Metallography
— Mechanical Properties — Summary — Bibliography

2. Ore Dressing   20

Comminution — Jaw Breakers — Gyratories — Secondary Crush-
ing — Cone Crushers — Grinding — Cylindrical Mills — Tube
Mill — Rod Mill — Ball Mill — Open and Closed Circuit Grinding
— Mechanical Rake Classifier — Flotation — Selective Flotation
— Flotation Machines — Pneumatic Cells — Sub-aeration Machines
— Reagents — Collectors — Frothers — Modifiers — Regulating
Agents — Activating Agents — Depressing Agents — General Re-
marks — Flotation in Practice — Gravity Concentration — Sluices
—Strakes — Tables — Jigs — Sink-and-Float — Electrical Con-
centration — Magnetic Concentration — Wet Magnetic Separation
— Belt Type Wet Separator — Electrostatic Concentration — Bib-
liography

3. Pyrometallurgy   55

Roasting — Heap Roasting — Stall Roasting — Hearth Roasting
— Cylindrical Furnaces — Blast Roasting (Sintering) — Smelting
— Blast Furnace — Reverbatory Furnace — Open-Hearth Reverba-
tory — Converter — Steel Converter — Copper Converter — Dust
Collection — Electrostatic Dust Precipitation — Refractories
— Carbon — Basic Refractories — Bibliography

4. Iron and Steel   75

Iron — Wrought Iron — Puddling Process — Mechanical Puddling
— Aston-Byers Process — Iron — Evolution of the Blast Furnace
— Blast-Furnace Reduction Process — Products — Pig Iron
— Slag — Gas — Dust — Recent Innovations — Tuyère Injection
— Steel — Cementation Process — Crucible Process — Sir Henry
Bessemer (1813-1898) — Thomas and Gilchrist — William Kelly
— Modern Practice — Bessemer — Basic Bessemer (or Thomas)
Process — Acid Process — Hot-Metal Mixers — The Integrated
Plant — Siemens Brothers — Open-Hearth Process — Electric
Steelmaking — The Arc Furnace — Current Regulation — Induc-
tion Furnaces — Vacuum Melting — Induction Vacuum Melting
— Vacuum Arc Melting — Use of Oxygen — Linz-Donawitz (L-D)
Process — LD/AC Process — Kaldo Process — Rotor — Oxygen
and Open-Hearth Furnaces — Alloy Steels — Stainless Steel — Bib-
liography

5. The Major Non-Ferrous Metals   127

Copper — Welsh Process of Copper Smelting — The Modern
Approach — Development of Copper Converting — Electrolytic
Refining — Electrolytic Production — Preparation of Alumina
— Bayer Process — Electrolytic Reduction — The Cell — Refining
— Compagnie A.F.C. Process — Aluminium Alloys — Wrought
Alloys — Cast Alloys — Output and Usage — Zinc — Smelting
— Horizontal Retorting — Vertical Retorting — Operation — Blast-
Furnace Smelting — Refining — Refining by Liquation — Refining
by Distillation — Operation — Electrolysis — Melting and Casting
— Galvanizing — Lead — Development of the Ore Hearth — Blast-
Furnace Smelting — Preparation of Feed — Roasting Furnaces
— Blast-Furnace Operation — Fire Refining — Preliminary Dross-
ing — Softening — Desilverization, Parkes Process — Dezincing
Operation — Recovery of Gold and Silver — Cupellation — Pat-
tinson Process — Electrolytic Refining — Nickel — Electrolytic
Nickel — Mond Nickel — Present-Day Nickel Extraction — Bes-
semer Conversion of Matte — Electrolytic Refining — Treatment of
Nickeliferous Iron Concentrate — Pressure Leaching — Tin — Ore
Dressing — Extraction of Tin — Reverbatory Smelting — Opera-
tion — Treatment of Slag — Refining — Electrolysis — Tin Plate

— Early Developments — Modern Tinning — Preparatory Treatment —Electro-tinning— Hot-Dip Tinning — Thickness and Weight of Coating — Bibliography

6. Newcomers in Metals   228

Uranium — Occurrence — Extraction of Uranium — Present-day Procedure —Alakaline Leach — Acid Leaching — Ion Exchange —Solvent Extraction — Preparation of Metal — Titanium — Arc Melting — Electrolysis — Alloys — Usage — Germanium — Earlier History — Zone Refining — Cobalt — Occurrence — Production of Smalt — Production of Metal — Arsenical Ores — Union Minière du Haut-Katanga — Treatment of Cobaltiferous Copper Oxide Concentrates — Rhokana Corporation Ltd., N. Rhodesia — Pressure Leach Treatment — Uses — Bibliography

7. The Precious Metals   266

Placer Gold — Amalgamation — Cyanide Process — Improvements in Sand Leaching — Refining — Chlorine Gas — Electrolytic Parting — Silver — Lixiviation Processes — Cyanidation — Argentiferous Base Metal Ores — The Platinum Metals — Recovery — Refining — Usage — Bibliography

8. Shaping of Metals   297

Shaping by Casting — Cast Iron — Gravity Die Casting — Pressure Die Casting — Investment Casting — Forging — Steam Hammer — Forging Press — Drop Forging or Stamping — Rolling — Extrusion — Wire Drawing — Powder Metallurgy — Bibliography

9. Metallography   321

Influence of Carbon — X-ray Crystallography — Radiography — Election Microscopy — American Metallographers — Bibliography

Glossary   332
Name Index   336
Subject Index   339

PREFACE

We study the past
Because it is a guide to the present
And a promise for the future.
The struggle for a better world is strengthened
By the hopes, ambitions, and deeds
Of those who were before us.
As we look backward
Our attention is directed forward.

A. B. WILDER

THE above lines from *A History of Steelmaking in the United States* are a reminder that even in the present age of rapid progress and change it is of advantage to take a look back to the past. History in addition to being a record of events is of interest in that trends in the past can often be used to help predict events in the future, and enable problems to be solved by clues furnished from earlier experience. It is in this light that the author attempts a backward look at Metallurgy.

This book is a record of the changes in metallurgy during the past hundred years. Following a conspectus it comprises eight chapters, each dealing with progress in one of the major branches of the metallurgical industry, the object being to select for description the most significant highlights; for it is impossible in a book of this size to attempt a comprehensive record of all metallurgical events during the period. The period covered by the text embraces the century 1850–1950, but the author has not hesitated to prior-date in the interests of continuity, also to record some earlier metallurgical aspects which have passed into oblivion but are entitled to historical record. Appropriate treatment has also been afforded to those developments which best illustrate recent progress made in metallurgical technology.

American metallurgical practice has been given prominence, for by reason of her immense mineral wealth, technical enterprise and abundant supplies of fuel the United States today is the foremost exponent in the extraction of metals. It is apt, however, to remember that although today physical metallurgy and metal performance form her main interest, Great Britain up to the mid-19th century was the largest producer of lead, copper, and tin, and that the metallurgy of these metals owes much to her pioneer efforts.

The references to literature from which a considerable amount of the material of the text has been drawn have been listed at the end of each chapter. The author wishes to acknowledge his indebtedness to the following for permission to make use of certain of their publications: British Iron and Steel Federation, Oxford University Press, McGraw-Hill Book Co. of New York, the American Institute of Mining and Metallurgical Engineers, United States Steel Corporation, Institution of Metallurgists, and the Sheet Metal Industries *Journal*. Almost inevitably some acknowledgements have been overlooked and for these I must tender my sincere apologies. I also wish to express my indebtedness to my wife for her valuable help in the preparation of the manuscript for publication.

W. H. DENNIS

Ilford, Essex, 1963.

CHAPTER ONE

INTRODUCTION

THE greatest purely metallurgical advance of the past hundred years or so was undoubtedly initiated by Sir Henry Bessemer, and William Kelly, a Kentucky ironmaster, who independently in the period 1850–55 discovered a process for steelmaking; for this invention ushered in the 'Age of Steel' in which we live. The importance of the discovery was that molten pig iron was transformed into steel within the space of some thirty minutes, a procedure that must be contrasted with the method previously in use for its merits to be recognized. Pig iron was first converted into wrought iron, which was then transformed into steel by heating in contact with charcoal for several hours in a cementation furnace. The iron took up carbon but not uniformly, carbon decreasing in quantity from the surface inwards. This heterogeneity gave rise to defects in the mechanical properties, being particularly troublesome to the makers of delicate apparatus. This lack of homogeneity was eventually overcome by Huntsman, a Doncaster clockmaker, who in 1740 succeeded in melting steel in a crucible and casting it. The merit of his invention consisted largely in the manufacture of crucibles which would stand up to the high temperature necessary to melt the metal.

The whole protracted process, involving as it did the production of wrought iron, its recarburization and then remelting in a crucible, occupied several days, and since the crucibles in the final operation held only 60–80 lb. of metal, output was never very large. Thus at one stroke the amount of metal handled by the new steelmaking process was far greater than had ever previously been contemplated. The first steel produced by Bessemer, however, was not of sound quality, the ingot (due to the presence of iron oxide which reacted with carbon) being full of blow holes rendering the metal brittle, fracturing being encountered in the subsequent working and forging. The trouble was eventually overcome by the addition of manganese in the form of spiegeleisen, a process which had been patented by R. F. Mushet just before Bessemer went into commercial production in 1856. Bessemer paid tribute to Mushet's help by seconding the proposal that the Bessemer gold medal of the Iron and Steel Institute should be given to him in 1876 for his spiegeleisen process.

Bessemer's discovery in 1856 was followed almost immediately by Siemens' open-hearth process in which oxidation of the metalloids in pig iron was induced not by oxygen in an air blast but by oxygen present in iron oxide which was added with the charge of pig iron.

In Bessemer's process the heat necessary to keep the metal in a liquid state is generated by the exothermic heat of the reaction, but in Siemens' process the heat of reaction was not evolved with sufficient speed to keep the metal in a molten condition, and heat had therefore to be supplied from an external source. This temperature was achieved in Siemens' furnaces by applying the principle of heat regeneration, the hot waste combustion gases from the furnace being caused to pass through a checkerwork brick chamber yielding up their heat to the checkerwork. Periodically the stream of hot waste gas was diverted to a second chamber, combustion air being then passed through the heated chamber. At regular intervals the direction of air and gas is reversed, so that each pair of regenerators alternated in absorbing heat from the spent gases and then imparting it to the incoming air and gas. Heat was originally initiated by the burning of solid fuel in a fireplace situated at each end of the furnace, but this was later displaced by the use of producer gas, which was also an invention of the Siemens brothers. As the process took up to ten hours as against thirty minutes with the Bessemer process, it was found possible to exercise far greater control over the steelmaking operation and to produce consistently a more homogenous, reliable and ductile material.

A further advantage of the Siemens method was that unlike the Bessemer method it can use a large proportion of steel scrap. By the late 1860s Britain was already heavily industrialized and large tonnages of cheap scrap were available. The advent of the open-hearth furnace meant that the scrap could be turned to good use.

Pig iron containing phosphorus was unsuited to the production of steel by the Bessemer process as originally developed, and as phosphorus is harmful to the properties of steel only those irons low in phosphorus could be used. Bessemer himself by the purest chance had used phosphorus-free iron in his original work. Although this difficulty at the time was not felt very acutely in this country, in Europe vast tonnages of high-phosphorus ores were excluded from conversion into steel. In 1878, however, a Welshman, S. Thomas, working with his cousin P. Gilchrist, realizing that the acid (silicious) lining was responsible for the non-absorption of the phosphorus in the (acid) slag, experimented with a basic slag. By relining the converter with dolomite (a basic refractory) and adding lime, a basic slag was produced capable of retaining the phosphorus present in the pig iron. This simple modification enormously extended the scope of steelmaking and was adopted in steel-making communities all over the world.

These three processes were the foundation on which the age of steel was erected. Their successful establishment brought many consequent developments which exerted a very wide influence on industry.

Alloy Steels. The steels available to industry in the late 1860s were high-carbon steels of the type made by the crucible process and low-carbon mild steels of the kind produced in the converter and open-hearth furnace. It was known that the hardness of steel depended on its carbon content; tools for instance, which were required to be very hard, being manufactured from high-carbon steels. The general employment, however, of such steel in industry was not possible owing to its brittleness. Thus in order to extend the usefulness of steel further, it was necessary to find other methods than raising the carbon content. The result was achieved by the introduction of certain non-ferrous metals into low-carbon steels, notably tungsten, manganese, nickel and chromium. One of the earliest alloy steels was introduced by R. F. Mushet who by adding tungsten to steel discovered self-hardening steel in 1868. Tools made by this method revolutionized machining processes, and it was also upon Mushet's self-hardening steel that the experiments were based which led to the production of the high-speed steels developed later in America.

In 1883 Robert Hadfield made an important step forward in

3

this field by incorporating manganese in steel. This alloy was found to possess remarkable tensile strength, elongation and hardness, and became invaluable for all machinery and plant subject to abrasive action such as railway crossings, dredger buckets and the like. These types of steel, however, did not provide a steel suitable for general constructional purposes, a start in this direction being made by J. Riley of Glasgow, who in 1889 by small additions of nickel to steel markedly increased the strength and toughness without decreasing the ductility. By addition of a further alloying element, chromium, H. Brearley in 1913 founded a class of constructional steels which in addition to strength and resistance to wear were also resistant to corrosion.

These alloy steels heralded in the Alloy Steel Age, and so great was their development that at the outbreak of the 1939 war there were no less than 2,000 different specifications dealing solely with alloys having various proportions of nickel chromium and small additions of other elements. With such developments as jet propulsion, nuclear fusion as a source of power and space technology, the acceleration in alloys is likely to continue.

Iron. Immediately prior to the invention of the Bessemer process, production of steel in Great Britain was of the order of 60,000 tons, whereas production of iron was 3 million. Iron was thus the principal metal on which the Industrial Revolution was founded. Iron and yet more iron, both wrought and cast, was the cry, and as the raw materials iron ore and coal were available in great quantity and in close proximity, the supply was adequate to the demand. Abraham Darby, a Shropshire ironmaster in the 18th century, by replacement of charcoal by coke enabled a much greater charge to be sustained in the blast furnace and thus made possible the production of pig and cast iron in much larger tonnages.

Henry Cort and J. Hall did the same for wrought iron, and Benjamin Huntsman produced the tool steel essential to the forming and shaping of castings, forgings, bars, etc. With these materials, and using the facilities which the newly discovered steam engine invented by James Watt in 1762 made available, extensive mechanization of industry took place. Essentially the Industrial Revolution was an engineers' revolution and not a metallurgical one. By the time, however, of the Great Exhibi-

4

tion of London in 1851, the metallurgical world was to experience a great advance in technical progress. Many factors contributed to this situation, new inventions, techniques and operational improvements, and a better understanding and appreciation of the scientific principle on which these operations are based. In addition the rapid developments in the use of iron and steel during the Industrial Age brought with them greatly increased demand for other metals, particularly copper, tin and lead. Moreover, the demand was not only for greater tonnages but also for a far greater variety of metals. Many of these metals were one hundred years ago little known names in the periodic table, but have now come into prominence and have become marketable commodities. It is accordingly not surprising that there have been more notable advances in metallurgy during the century under review than in the whole history of this ancient art.

Non-Ferrous Metallurgy. The above outline has briefly enumerated some of the features of iron and steel manufacture during the past century. Developments in the non-ferrous field were likewise proceeding on every front.

Production of copper and lead received early impetus following the introduction of the dynamo in the second half of the 19th century; in fact the electrical industry was destined to become the greatest customer for these two metals and to a lesser extent practically all other non-ferrous metals. Later on, the automotive and aviation industries were to constitute the principal outlet for the light metals aluminium and magnesium. The phenomenal rise in the production of aluminium since its first commercial production in 1888 has not been matched by that of any other metal. World production today at 5 million tons is greater than that of any metal other than iron and steel. Its valued properties of lightness, corrosion resistance, electrical conductivity combined with moderate cost have extended its use into a wide variety of commercial and domestic fields.

The methods by which lead, tin and silver were extracted from their ores are among the earliest of the ancient arts. That there was an important trade in these materials in Great Britain is undoubted and it is commonly held that the Phoenicians came to Cornwall for their supplies of tin. We know also that the Romans smelted galena from the Mendip Hills and elsewhere for their production of lead. In spite of the fact that our country owed its early development to the exploitation of minerals, it is a surprise to most people that Britain in the mid-19th century was dominant in the production of the non-ferrous metals. For more than 150 years during the 18th and 19th centuries, Swansea was the biggest copper smelting centre in the world. At the height of the boom, Swansea docks were so packed with copper ore ships from all parts of the world, that it was possible to walk from dock to dock by stepping from one ship to another.

Tin, one of the oldest established of British exports, was being mined and smelted on a considerable scale from Cornish ores. In the smelting of lead this country had been pre-eminent since the days of the Roman occupation. Zinc smelting was firmly established in Bristol and Swansea.

During the second half of the 19th century, however, Britain lost her pre-eminence, the copper and tin smelting industries are now fading history; only a few clusters of ruined copper works testify to the former greatness of Swansea. The chief reason was the development of natural mineral resources and progressive industrialization abroad. Britain had never been a richly mineralized country, the domestic mines were being worked out, the copper, lead and zinc smelters being no longer able to purchase supplies of rich ores from overseas mines in competition with smelters on the mine site itself.

America especially was endowed with abundant resources of metal ores, large deposits of copper ore being especially prolific, a large copper smelting industry being built up between 1850 and 1900. Similarly in South America, the porphyry copper ores of the western slopes of the Andes in Chile and Peru were being developed. Australia discovered that she was well endowed with zinc and lead ores, and laid the foundation of a large smelting industry in these metals.

Alluvial tin was found in the East Indies and being much more accessible than the lode tin of Cornwall, eventually sounded the death-knell to British tin mining. As a consequence Britain became an importing country for the non-ferrous metals she required and today there are few non-ferrous smelters of any size in this country. The zinc smelters at Avonmouth and Swansea depend for their raw material on imports from abroad, as does the Mond Nickel Co.'s plant at Clydach, a few miles from Swansea, which recovers nickel metal from matte imported from Canada. The only other smelting concern of any size is that of the British Aluminium Company in Scotland which imports its raw material—bauxite. There is now no large plant in the U.K. for the production of lead, nor is there a plant for the smelting of copper ore, which must be regretted.

Ore Dressing. Unlike ferrous metallurgy, in which the smelting operations are applied to the crushed, screened and sintered ore, the first step in non-ferrous metallurgy is beneficiation, in which waste is removed, concentrating the valuable minerals into a smaller bulk for the subsequent costly smelting treatment. Today beneficiation is of far greater importance than a century ago, for the rich ores of those days only necessitated a crude hand picking for either selecting the rich ore or rejecting the barren waste. This was sometimes supplemented by gravity methods. Preferential separation, however, is seldom possible with gravity concentration, so the concentrates were usually of only moderate grade and sometimes so complex that they were not amenable to the treatments that were then available. The introduction of new methods was therefore an advance of the greatest importance, for they rendered profitable not only low grade deposits, but a number of rich complex ores that had previously been unworkable.

Of the new beneficiation methods, the flotation process which was introduced at the turn of the century has undoubtedly been of the greatest importance, in fact it has been termed 'one of the greatest process discoveries ever made for the betterment of our standard of living'.

No detailed description will be given here (see page 28); it suffices to say that active research on flotation began just before 1900 and led to the process of bulk froth flotation which resulted in the production of a concentrate containing substantially all the sulphide minerals. This limited the process to the beneficiation of ores in which only one valuable sulphide mineral existed, since bulk flotation was incapable of dealing with complex ores. Later, however, in the early 1920s it was found possible to separate the sulphides of the various metals present in the ore by a selective flotation operation and it is now possible to separate three or four different metal sulphides from the same ore. The process has become very important in the beneficiation of the major non-ferrous metals, copper, zinc, nickel and lead.

The selective process has also made possible the treatment of complex ores that formerly were of little value due to the costly metallurgical requirements for the extraction of the metals.

The effect on the economy has been an important one, for the flotation process has been a vital factor in increasing the world ore reserves by making it economical to work lower grade ores and to use the leaner portion of the orebodies which were formerly left in the slopes.

Not only in metallic ore beneficiation has flotation become important but also in non-metallic material, for it is now being used in the treatment of phosphate rock, fluorspar, barytes, feldspar, etc.

Flotation and many of the other beneficiation methods produce material in a very fine state of subdivision. The blast furnaces and some other operations are incapable of treating such material, lump ore being required for feed material. Hence such methods as sintering, pelletizing, and briquetting have been developed. Of these, sintering is the most frequent method in use. In this process, air is drawn through a moving layer of the material which has been previously ignited, the air supporting the combustion of sulphur present in the ore (or added carbonaceous material such as coke dust). The high temperature causes the charge to frit or fuse together into compact lumps. This process has become of particular importance in the iron industry where the ores now obtainable have become finer and finer. The sintered iron ore is in a physical condition extremely favourable for reduction in the blast furnace, and nowadays almost all iron plants have sintering installations.

Pyrometallurgy. Smelting usually necessitates a dry coarse feed with a low sulphur content, and in roasting, which is used principally for lowering the sulphur content of ores, progress has mainly been in the improvement of mechanical equipment and not in chemical fundamentals. Reverberatory roasting furnaces with fixed hearths necessitating hand-rabbling, expensive in labour and fuel, which were in use in the mid-19th century, gave way to the mechanical-rabbled type (Edwards, Merton, Ropp, etc.) and finally to the mechanical-rabbled multiple-hearth furnace as exemplified by the Wedge and

Herreshoff. The original 7-hearth water-cooled McDougal wedge multiple-hearth furnace has been gradually developed to 8, 12 and up to 16-hearth, 100 tons of sulphide material per day being roasted with practically no fuel.

During recent years the fluo-solid roaster has been introduced and in certain fields is seriously challenging the multiple-hearth roaster. In this process the finely divided feed is held in suspension by a stream of gas, combustion having been initiated by a burner. It requires simple equipment, does away with rabbling, gives intimate contact between air and solid resulting in high thermal efficiency.

One hundred years ago carbon reduction was prominent in the smelting of lead, tin and zinc and is still the most commonly used procedure. The blast furnace still dominates in iron and lead smelting; but is now used very little in the copper and nickel industry, reverberatory smelting having taken over almost all smelting operations. This furnace has been brought to a high state of perfection and efficiency from the old hand-operated 13 ft. long copper furnaces expensive of fuel and labour. It possesses the advantage over the blast furnace that a variety of fuels can be used, pulverised coal, gas, oil, tar and pitch, most of which are cheaper and more convenient to handle than coke. The blast furnace has, however, recently received a fillip, for a recent publication* states that after 25 years of study and research production of zinc in a blast furnace is now possible. As lead can also be recovered, and as lead and zinc so frequently occur in intimate association in their ores, the importance of this development is evident.

Electrometallurgy. In 1831 Michael Faraday made his vital discovery of the relationship between magnetism and electricity, from which flowed the invention of the dynamo and the electric motor. The benefits resulting from these inventions, however, were obtainable only where electricity was available in bulk, which did not occur until some forty years later—well into the second half of the 19th century. Electricity had a many-sided impact on metallurgy. Copper, for instance, because of its high electrical conductivity and the ease with which it could be drawn into wire became in great demand as an electrical

* S. W. K. Morgan, 'Production of Zinc in a Blast Furnace', *Bulletin I.M.M.*, August 1957.

conductor, and thus copper made possible an enormous expansion in electrical generating and other equipment. On the other hand, the electrical generators made possible the electrolytic refining procedure necessary to produce copper of the required high standard of purity. The electrical industry also became the major customer for lead and a great consumer of other metals.

The commercial utilization of electrolysis in metallurgy had its origin nearly 100 years ago when James Elkington, an English electro plater, invented a process for refining copper electrically; later, about 1890, aluminium was first produced on a commercial scale by electrolysis, by Hall in America and Heroult in France, followed by lead in 1903, nickel about 1910 and zinc in 1915.

In addition to providing a means of obtaining a metal direct from its ore it also serves to purify a metal produced by some other process. At the end of the last century it was limited to the refining of gold, silver and copper, and to the production of a small amount of aluminium. Since that time, most of the copper, nickel and magnesium and a large part of the zinc and lead are produced electrolytically, cathodes of copper, nickel, lead, zinc being made close to 100% pure.

Electrolytic manganese and cobalt have also in recent years been introduced, as have many of the rarer metals such as beryllium, indium, tantalum and tellurium.

Electrolytic methods using fused salts as electrolytes are also of the greatest importance; these are used not only for the production and refining of aluminium, but also for the winning of magnesium, and may in the future be employed for the production of titanium.

Electrolysis also possesses another valuable attribute in that during the refining process precious metals such as gold, silver and platinum associated with the metal being refined are deposited in the cell as a slime, their recovery constituting a valuable by-product.

Electric Smelting. The development of cheap power towards the end of last century enabled the electric current to be utilized for heating purposes. Very high temperatures were attainable by its use which is important in the production of such materials as ferro-alloys. It has also the advantage that the furnace atmosphere can be made reducing, oxidizing or neutral at the will of the operator; further a product of higher purity can be obtained than by the use of other processes using conventional fuel. Electric pig iron, electric steel and steel alloys generally have the advantage of high quality. The arc furnace was introduced into the steel industry at the beginning of this century and originally used for the manufacture of tool steels. It was extensively used during World War One for the production of alloy steels for ordnance purposes and at present it is commonly used for stainless and manganese steel as well as the whole range of low alloy steels for the automobile and aircraft industry. In countries possessing cheap hydro-electric power such as Italy and Scandinavia the arc furnace has also for many years been used for making ordinary quality carbon steels. This use is now extending to other countries, furnaces having been built in America, Europe and in Britain. It is likely that such furnaces will in the future actively compete with the basic open hearth as an economic producer of ordinary quality steel.

High-Frequency Induction Melting introduced during the 1920s largely replaced the crucible furnace as a melting unit for tool steel and high alloy steels. In connection with the latter the induction furnace was one of the first to be used for vacuum melting, which overcomes the unfavourable effect of the atmospheric gases on the mechanical properties of the steel. A new development in vacuum melting is the melting of such reactive metals as titanium, zirconium and other rare metals in the consumable electrode vacuum arc furnace using a water-cooled copper crucible. It has enabled ingots of the reactive metal to be cast in sizes of up to 5 tons in weight. Consumable electrode vacuum melted steels up to 20 tons weight are also being produced. These steels possess a high degree of cleanliness, decreased gas content, soundness, uniformity of ingot structure, enhanced toughness and ductility at elevated and sub-zero temperatures. These are the properties which provide the engineer with the quality materials of the new technological era.

Hydrometallurgy is in general the process of recovering a metal by a solvent and has found application in the treatment of certain types of ore. Examples of such processes are the leaching of copper ores, the cyanidation of gold and silver ores and the Bayer process for extraction of alumina from bauxite.

The first large scale application of the method occurred in the 18th century at the property of the Rio Tinto copper mine in Spain. Large heaps of the crushed sulphide copper ores were allowed to oxidize over long periods under the influence of moisture and air, with subsequent leaching out of the copper sulphate that had formed from the oxidation of the sulphides, followed by precipitation of the copper on scrap iron. This simple method has also been applied elsewhere, notably in America. Confined leaching, i.e. treatment in tanks as opposed to heap leaching, has been practised since the beginning of this century, large scale operations having occurred at the plant of the Chile Copper Company at Chuquicamata, Chile (at a rate of 1 million tons per month), New Cornelia Copper Company at Ajo, Arizona, and the Union-Minière in the Congo.

In recent years there has been a marked extension in the application of hydrometallurgical processes to low grade ores due largely to the introduction of new techniques such as ion-exchange, solvent extraction and high temperature and high pressure methods. Ion-exchange used for the softening of water has found employment in the purification and enrichment of the solution resulting from the acid leach of uranium ore. Solvent extraction widely employed in the petroleum industry is now used in the separation and recovery of uranium, tantalum, etc. High pressure and high temperatures have long been utilized in the chemical industry but it is only comparatively recently that their usefulness has been recognized in connection with hydrometallurgical processes. The first plant employing high pressure and temperature was initiated in Canada at Fort Saskatchewan in 1953 for the treatment of nickel-copper-cobalt sulphide ore, and it has since been followed by similar plants in America and Cuba.

Shaping of Metals. The last stage in smelting usually involves the casting of the metal into a mould. This mould may be shaped to the form desired in the finished article, the process being known as founding or casting. On the other hand the metal may be cast into a mould of simple form such as an ingot for subsequent shaping by such mechanical working methods as forging, rolling, extrusion, etc.

Casting. The last hundred years has seen marked improvements and innovations in this ancient art. Pressure die casting involving the injection of metal into metal dies or moulds under pressure has expanded greatly since World War One. It is the fastest of all casting processes, a large tonnage of aluminium and zinc base die-cast alloys being produced for the automotive, aircraft and engineering industries. Investment casting has also been modernized and introduced as a mass production technique.

The application of continuous methods to the casting of ingots has made rapid strides since the last war. This method of casting involves the continuous solidification of the ingot while it is being poured, the length not being determined by mould dimension. Liquid metal enters one end of a mould continuously, solidified metal emerging from the other end in a long length of the required cross-section which can be cut to length for further processing. Casting by this method eliminates the casting bay, ingot moulds, soaking pit and primary rolling mills. In addition to the saving on capital and reheating costs there is usually an improvement in the yield of saleable product as compared with conventional casting and rolling.

As far back as 1856 Henry Bessemer realized the advantages of the technique and in fact produced sheet by pouring molten steel between a pair of water-cooled rolls, but the product was lacking in quality. During the intervening hundred years many different ideas for the direct casting of liquid metal continuously to semi-finished product were suggested and patented.

Continuous casting machines first achieved success in the non-ferrous field particularly for copper, brass and aluminium. Progress with steel was negligible until the last war, when notable developments were made in Germany, Russia, the United States and the U.K.

Forging. The introduction of the steam hammer, invented by J. Nasmyth in 1842, occurred at an appropriate moment, for it enabled the heavy masses of steel produced a few years later by the Bessemer and open-hearth processes to be readily forged and shaped by rapid and repeated blows dealt by the hammer. Formerly, forging was achieved by the tilt hammer, a crude contrivance consisting of an iron hammer head fixed at the end of a long wooden beam, pivoted at the other end and actuated by a steam driven cam, which raised the hammer, which then fell by its own weight on the mass of hot metal on the anvil. This type of mechanism would have been inadequate to forge the heavy masses of Bessemer metal.

For very large forgings the hydraulic press was developed by Sir Joseph Whitworth in 1856, the ingot being shaped not by impact but by squeezing induced by pressure from a hydraulic ram. All heavy forgings such as boiler drums and big guns are nowadays formed by pressing.

Rolling. For making long and thin sections of metal, such as sheet and plate, reduction in thickness was originally secured by pounding the metal under a water-powered hammer. For soft metals such as gold and silver this was a fairly simple operation, but for hard, tough metals the method was laborious. Rolling mills driven by water wheels were introduced towards the end of the 17th century, John Hanburg in 1697 erecting a rolling mill at Pontypool, S. Wales, for rolling iron for making tinplate. In Sweden bars and sheets of wrought iron were being rolled on a mill invented by C. Polheim (1661–1751). The employment of rolling mills only began extensively late in the 18th century owing largely to the work of Henry Cort in 1783, who introduced and patented the use of grooved rolls for producing bars from wrought iron. The demand for steel in a greater variety of forms led later to the development of differently shaped grooves for the production of complex shapes and sections.

The rolling mill reached a high standard of development towards the end of the 19th century when steel rollers of sufficient size and strength could be cast and when ample power became available. It is now the most widely used of all metal shaping processes, for steel and non-ferrous metals are rolled in some if not in all stages of fabrication. The growth of mills for production of continuous strip and cold reduced sheet in quantity and quality has been the result of an insatiable demand from old and new industries and is clearly reflected in the enormous output of strip, plate, sheet, bars, wire, structural shapes and foil.

It can be stated that in no other phase of the fabrication of metals has mechanical ingenuity and metallurgical knowledge been so clearly evidenced as in the construction and operation of the modern rolling mill.

Extrusion. The method of producing special sections by forcing heated metal billets through a suitably shaped die was developed early in the last century. It was at first confined to the softer metals such as lead, but was later extended to brass, and then aluminium, magnesium and their alloys. Within the last few years the method has been applied to steel.

Metallography. In 1861 Professor H. C. Sorby of Sheffield initiated the systematic examination of metals through the microscope and laid the foundation of that branch of metallurgy known as metallography. Sorby's work was largely confined to iron and steel and was accomplished during the years 1860–65. He had previously worked on the microscopical structure of rocks which could be examined microscopically by transmitted light. Metals being opaque were not amenable to such a method, but Sorby was successful in devising suitable techniques using reflected light; his publications contain photomicrographs which illustrate the excellence of his technique. His early observations remained almost unnoticed, until 1887, when he first published his findings in a paper to the Iron and Steel Institute.

Further microscopical work was carried out, mostly with the intention of controlling the quality and composition of iron and steel, in 1878 by A. Martens in Germany, in France by F. Osmond in 1880, and in England by Roberts-Austen.

The constituents of the microstructure of a metal are formed during solidification from the liquid; hence the appearance of the structure alone will not reveal information regarding its origin or the relation of one structure to another. For this the microscope has to be supplemented by other instruments, notably the pyrometer. With its aid thermal equilibrium diagrams were constructed that depict the phases present at all temperatures between room temperature and that at which the metal or alloy melts. This type of work was followed up in many countries, notably France, and as a result metallurgists were able to indicate the structure which was likely to occur in any particular composition of alloy, when (*a*) cast, (*b*) quenched from different temperatures, or (*c*) tempered, and further that the mechanical properties were related to and dependent upon structure.

These diagrams could be put to practical use, for they indicated the nature of the heat treatments that would produce a desired result, and it is through the data contained in the thermal equilibrium diagrams that methods of heat treatment for different metals and alloys were made possible.

Following on the discovery of X-rays by Röntgen in 1895 the commencement of this century saw the application of X-rays to elucidate the internal structure of metals and alloys and the foundation of a new science was laid. From this work many discoveries of the arrangement of the atoms in metallic crystals have emerged, leading to a better understanding of alloy structure. The invention of the electron microscope, with which magnification may be obtained many hundreds of times greater than was possible in Sorby's time, has greatly facilitated the observation of internal structure. Primarily many of these investigations are the work of the physicist and their connection with metallurgy is not easy to recognize.

Mechanical Properties. A century ago structural design of engineering components was based almost entirely on such static properties as tensile strength and hardness. At that time sources of motive power were limited and the design based on static strength properties provided an ample factor of safety.*

With the development of the steam engine and a higher degree of industrial mechanization, rupture of materials began to be increasingly experienced, which could not at that time be accounted for, as the ruptures occurred at stresses well below those which the metal was capable of withstanding under static load conditions. Fracture of the crankshafts of engines and axles of locomotives and wagons were but two examples. Investigation by Sir William Fairbairn in the early 1860s and by August Wöhler a German engineer a few years later showed that if metals are subjected to variation in load, wide fluctuation in stress occurs which if of sufficient magnitude results in fracture at stresses well below the maximum static strength. The term 'fatigue' was applied to such fractures because they occur only after a prolonged period of service. The term was appropriate

* Factor of safety is the ratio allowed in design between the breaking load on a structure and the safe permissible load on it.

16

in that under repeated stress the capacity of a metal to withstand stress gradually diminishes. It is, however, inapt in that metals do not regain their original properties after a period of rest.

Fatigue has become progressively more prevalent in modern times as technology has given rise to more equipment subject to repeated loading and vibration, such as automobiles, aircraft, steam turbines, etc. All rotating machine parts such as shafts, axles, piston rods, etc., are subject to cyclic stresses. Vibration due to machining is also a frequent cause of failure. It is particularly invidious because fractures occur without any obvious symptoms or warning. The phenomenon was highlighted in modern times by the 1954 disaster to the original British Comet jets produced by fatigue failure at a hatchway as a result of alternating changes in the cabin pressure with altitude.

Today it has been stated that fatigue accounts for at least 90% of all service failure due to mechanical causes. On this account the fatigue strength of a metal is today one of its most important mechanical properties for in all cases of dynamic loading at normal temperature it is directly or indirectly the limiting factor of design.

In many applications materials are subjected not to cyclic (fatigue) stresses but to steady static loads for long periods of time, e.g. steel cables, furnace and steam boiler equipment, blades of turbine and jet engines. Under such conditions the material may gradually elongate until its usefulness is seriously impaired. This mode of deformation is known as 'creep', so-called because the phenomenon occurs at a slow rate over a relatively long period of time.

Creep will take place in every metal or alloy if it is subjected to stress while maintained above a certain minimum temperature. The majority of metals do not, however, exhibit the phenomenon of creep to any noticeable degree except on the application of heat. Initially, interest in the subject centred around those steels used in steam power plants, oil refineries and chemical plants in which the operating temperature seldom exceeded 550° C. In such employments the components were comparatively only lightly stressed, but during and since World War Two, developments in many fields have necessitated materials possessing stability of physical properties and dimensions at high stresses and temperatures. The high temperatures and pressures encountered in such applications as the super-heating of steam, the cracking of oil, the jet engine, guided missiles, nuclear power plants, space technology, all call for materials with good high temperature and creep-resistant characteristics. In gas and steam turbines the dangerous aspect of creep is in the loss of clearance of the moving blades, for any failure in this respect can lead to disaster.

The changes with temperature of the resistance of metals to fluctuating stresses (fatigue) are not essentially different from the changes of their resistance to constant stresses (creep); hence alloys developed for high creep strength may also be expected to have high fatigue strength.

Much effort has been directed towards the development of new structural metals such as titanium, vanadium and niobium, and their alloys, by reason of their creep strength, but the range of materials available rapidly decreases as the temperature level rises; in fact above 1000° C. existing alloys are in many cases inadequate for the environment and stresses imposed. This fact has prompted research into such unconventional materials as for example the cermets (an abbreviation of the words ceramic and metals) designed to combine the desirable properties of metals with the refractory heat characteristics of the ceramics.

Summary. How short a distance history had advanced along the road towards the present may be gauged by the fact that in the mid-19th century the only metals that were available to industry were cast iron, wrought iron, a small proportion of tool steel, and of the non-ferrous metals, only copper, zinc, lead, tin, and the alloys of these metals—brass, bronze, gun-metal, pewter and solder. Since then progress and advances made in metallurgical technology during the period under review have enabled metal art to be replaced with metal science. Certainly no similar period in the history of metallurgy compares with the progress recorded here.

In production alone, world metal output during the past half-century has been greater than in all preceding history. Not only have old methods been improved and expanded to meet the increased demand for metals that were in use one hundred years ago, but metals that in 1850 were mere laboratory curiosities are today being produced by the ton and form the basis for alloys and structural materials demanded by the radical new environments in which man is now operating. The advent of nuclear energy, guided missiles, jet engines and space technology has necessarily led to the creation of new tools, techniques and materials.

BIBLIOGRAPHY

1. AITCHISON, L. (a) *History of Metals*, MacDonald & Evans, London. (b) 'A Hundred years of Metallurgy, 1851–1951', *Sheet Metal Industries Journal*, May and June 1951.
2. 'Seventy-five years of progress in the mineral industries' (1871–1946). American Institute of Mining and Metallurgical Engineers, New York.
3. DANNATT, C. W., 'Fifty Years of Extractive Metallurgy', *Mining Magazine*, January 1960.

CHAPTER TWO

ORE DRESSING

ORE, the product of the mine, consists of mineral or minerals sparsely disseminated through barren rock material known as gangue. Elimination of as much as possible of this rock material is an economic necessity before smelting, for (1) the rejected material is not transported and hence a saving in freight costs is achieved; (2) a mechanical rejection of gangue is much cheaper than elimination of waste material as slag in a smelting process. The operations involved in the elimination of the gangue and the concentration of the mineral(s) into a small bulk is known as mineral dressing. Several other terms are also in common use, namely ore treatment, ore dressing, milling and beneficiation. The valuable product of the operation is known as concentrate and constitutes the feed to the metallurgical plant; the discarded waste is called tailing.

As the mineral(s) occur intimately associated with waste rock, before any separation and concentration can take place the minerals must be liberated from the gangue, and hence severance or comminution constitutes the first step in mineral dressing. After liberation the mineral(s) are separated from the gangue and collected into an enriched concentrate. It may be stated that mineral dressing does not effect 100% separation of the mineral value from the gangue, and in fact in nearly all cases the mineral concentrate obtained usually contains at least 10% barren material.

One hundred years ago the progress in the art can be seen from the accompanying flowsheet (Fig. 1) of the Clausthal mill (Germany) which was one of the largest and most up to date at that time. The ore which was galena and zinc blende associated with copper and iron pyrites was hand-sorted, crushed and sized, the coarse sizes being jigged and the fines subjected to buddles. The flowsheet should be compared to that on page 39, which shows the modern approach to beneficiation of the same type of ore.

Comminution. The ore as mined comprises lumps up to 4 ft. and hence reduction in size must be effected in order to break down the ore and release the mineral values. Since no single

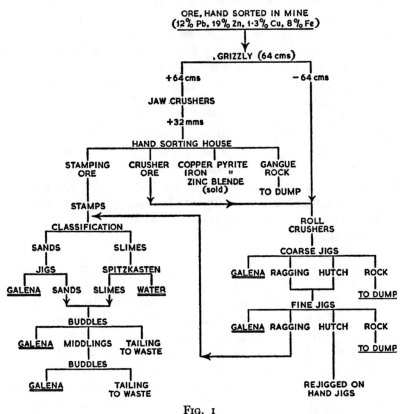

FIG. 1

FLOWSHEET SHOWING LEAD RECOVERY AT CLAUSTHAL MINE (GERMANY) IN 1871 (capacity 500 tons per day)

RANDOLPH, A.I.M.E. 1877, 6, 470

machine is capable of effecting the entire crushing operation, it is customary to conduct the operation in stages, the lump ore being first broken down in primary crushing machines, further reduction being effected in secondary crushers which reduce the ore to $\frac{1}{2}$ – 1 in. Primary crushing is usually effected by means of jaw crushers or gyratories.

Jaw Breakers.   The first successful mechanical rock breaker was invented by Eli. W. Blake (1795–1886) of New Haven, U.S.A., and patented in 1858, being first exhibited in Great Britain in 1862 at the London International Exhibition. Blake adopted the principle of the toggle linkage, the rock being broken between a horizontal fixed jaw plate and a movable jaw plate, hinged at the top and set at a small angle and actuated by a powerful toggle movement communicated to it from a driving pulley through an eccentric shaft and pitman which is made to move up and down. The movable jaw is held up against the toggles by a tension rod and spring, advancing and receding a short distance alternately from the fixed jaw. Rock is fed in at the top between the two jaws and is broken by the pressure exerted by the moving jaw, gradually sliding down until it is finally crushed sufficiently to fall through the opening at the bottom of the jaws. A jaw crusher 30 in. $\times$ 15 in. when operating on an average product of say $-$ 8 in. $+$ 2 in. has a duty of about 40 tons per hour. The advantages of simple rugged construction and large capacity are such that today the Blake crusher is still the leading machine for heavy duty primary crushing.

Gyratories.   The next major advance in crushing was the utilization of the gyratory principle embodied in the gyratory crusher and its outgrowth the cone crusher. In 1881 P. W. Gates in America was granted a patent for a machine which included in its design all the essential features of the modern gyratory crusher. The machine consists of a central vertical shaft which carries at its upper end a conical crushing head gyrating within a stationary crushing surface in the form of an inverted cone. The central movable crushing head receives impetus via an eccentric from a driving pulley and gyrates around the periphery of the fixed crushing head. Ore is fed in at the top and crushed between the surfaces of the breaking head and that of the fixed crushing face as the former advances towards the latter, the crushed material falling through as it recedes, the same function being performed as the movable jaw in the jaw crusher. The modern gyratory runs at about 500 r.p.m. the capacity depending on the size, a machine capable of dealing with rock up to 8 in. in diameter having a capacity of 60/70 tons per hour, of which about 60% is $-$ 1 in. Large machines which take up to 12 in. rock have a capacity of 150 tons per hour.

Secondary Crushing. The size of product from primary crushers varies from about 2 in. up to 6 in. or so according to the size of machine, and as concentration processes call for crushing down to $\frac{1}{2}$ in. or finer, the need arose for secondary crushing. This requirement was met for a number of years by the employment of crushing rolls formerly used as hand-operated machines on the small lead mines of Derbyshire, a mechanical type being introduced by John Taylor in 1804 at the Wheal Crowndale mine in Devon. These consisted of two iron cylinders with their axes placed horizontally and so mounted on shafts that they revolved at a distance of about $\frac{1}{4}$–$\frac{1}{2}$ in. apart according to the fineness of the product required. Rock is fed between them and broken by the pinching action produced. The early rolls suffered from many disadvantages, constant attention being required to keep them operating efficiently. Corrugations on the roll surface soon occurred which affected the uniformity of product, repair costs being high. S. R. Kron of Jersey City in 1874 much improved the rolls by making them with hard renewable shells enabling a longer life to be attained. At the present time rolls up to 6 ft. diameter are in use, being specially favoured in the crushing of clayey iron ores for they can handle sticky feeds better than most other forms of crusher.

Cone Crushers. The earliest attempts to adopt the gyratory principle to fine crushing took place at the beginning of this century, but it was not until after World War One that gyratory crushing as exemplified by the cone crusher entered the fine crushing field. In its adoption to fine crushing the gyratory crushing head was made much more obtusely conical, the additional flare resulting in a head of much greater diameter with a consequent large increase in the area of discharge opening. This was essential in order to maintain capacity in face of the reduction in width of discharge opening incidental to finer crushing. In order to overcome the tendency of the material to slow down in speed as a consequence of the increased flare of the crushing head, the eccentric speed was increased, thereby increasing the number of crushing impulses per unit area of crushing surface per unit time. Increase of the movement of the head was also adjusted by giving a longer throw. The result of these modifications was to ensure faster travel through the crushing zone. Originally these machines were fitted with straight face concave crushing surfaces; later the concaves were tapered at both ends to distribute the wear better. Eventually these non-choking concaves became standard equipment. The cone crushers proved popular from the start and are now standard equipment in practically all ore and quarry comminution plants.

Grinding. Grinding is a process of pulverizing ore or other material by a combination of impact and abrasion. Grinding is similar to crushing in that breakage by impact is common to both, but the essential difference is that a much smaller size of particle is secured by intimate contact of the crushing surfaces, reduction thereby being achieved by rubbing or abrasion.

The primary purpose of grinding is to liberate the valuable minerals from the gangue, and hence some form of grinding unit is almost universal. These in modern practice take the form of tumbling mills employing balls, pebbles or rods as the tumbling or grinding media. Grinding in water is usually employed, but in a recent development (Aerofall mill) dry grinding is used.

CYLINDRICAL MILLS

The machines most commonly in use for performing fine grinding are cylindrical-type mills mounted horizontally on their axes. There are several types, differentiation being on the basis of the tumbling media used inside the mill to effect the grinding. Ball mills employ steel balls as the tumbling bodies, rod mills steel rods, and tube mills flint pebbles. A ball mill also differs from a tube mill in that the length is usually not more than $1\frac{1}{2}$ times the diameter, whereas in a tube mill, the length is much greater than the diameter. The approach to the modern grinding mill may be discerned in the invention (by F. W. Mitchell and H. T. Tregoning of Cornwall) of what was termed a barrel pulverizer in 1880. This consisted essentially of a cast iron cylinder mounted on horizontal bearings. Pieces of iron and steel were charged to the mill, which was rotated at 8 r.p.m. Ore was introduced into the mill through one of the hollow bearings with a stream of water, the ground product flowing out at the other end.

Tube Mill. Following on this early development of the pulverizer came the tube mill first used in the cement industry. The mill consists of a steel cylinder about $5\frac{1}{2}$ ft. diameter by 22 ft. long lined with flint blocks and rotated on hollow trunnions. The mill was loaded with pebbles (10–12 tons) and rotated at a speed of 28 r.p.m., the capacity measured in terms of tonnage of 100 M product being about 150 tons per day. In the early days a peripheral discharge was used, screens being arranged around the periphery at the discharge end of the mill; but excessive wear caused its abandonment in favour of central trunnion discharge. Composite steel and pebble loads are now finding favour, an increase of about 15% in capacity being obtained. The tube mill received a big impetus on its introduction to the Rand goldfields in 1904 by J. R. Williams, for the purpose of grinding the ore leaving the stamp mills to achieve the degree of fineness required for the subsequent cyanide processes which had been introduced some time previously.

Rod Mill. Substitution of steel rods for balls as grinding media characterizes the rod mill introduced from Germany at the turn of the century. Since the action of the rods resembles that of a large number of close-set rolls it is to be expected that the product would be roughly similar, and in fact the size distribution is reasonably so. Like rolls the rod mill is most suitable when a granular product is desired with a minimum of fines, and hence found use in gravity concentration plants, being particularly adapted to tabling.

Ball Mill. A mill with its diameter greater than its length was developed by H. Gruson of Magdeburg in 1885. It was subsequently taken over by Fried. Krupp AG and known as the Krupp mill. It consisted of a steel drum whose circumference was composed of fine screen material. Within these screens were coarse screens made of perforated steel plate so arranged as to protect the outer fine screens. Inside the protecting screens are iron or steel perforated plates, so arranged in helical form as to constitute a series of steps, which has the effect of raising the balls and then dropping them during the rotation of the drum. In operation feed is introduced from a hopper at one end of the cylinder and is crushed by the impact of the balls against the grinding surfaces of the lining and against each other. When sufficiently ground the ore passes the large apertures in the grinding plates on to the surrounding coarse screens which intercept the larger particles whilst permitting the passage of fine material through to the inner screens. This discharges into a hopper which forms part of an outer casing of sheet iron or steel surrounding the rotating drum. The material that does not pass through the screen falls back into the interior of the crushing chamber through openings between the lining plates. A 5 ft. diameter by $3\frac{1}{2}$ ft. wide mill revolving at a speed of 26 r.p.m. has a dry capacity of about 25 tons of 60 M material per 24 hours. Charge of balls is 2,000 lb. of 4 in. diameter.

A nearer approach to the modern ball mill was the Grondal, introduced into Scandinavia in the 1890s for the wet grinding of iron ores. It was commonly 6–7 ft. in diameter by 3–4 ft. wide and made up of longitudinal steel bars fastened with cast iron end plates protected by liners. The mill was mounted horizontally on trunnion bearings although in some cases tyre and rollers were employed. Cast iron balls up to 6 inches were charged until about half full. Feed was introduced through one of the trunnions, the ground product flowing out of the other. As the balls wore down, others were added to compensate for the wear, the average wear being about 2 lb. of metal per ton of ore crushed. The mill was run at 26–28 r.p.m. and crushed about 50 tons of ore per day down to 30 M.

H. Hardinge in America in 1908 designed a ball mill which instead of being cylindrical was of conical shape, the cone construction causing the balls to segregate roughly according to size, the larger balls in the large diameter, the small balls segregating at the discharge end. Thus the coarse feed on entering the mill is reduced by the larger balls in the section of greatest diameter, the balls falling from the maximum height which ensures the greater force being exerted on the pieces hardest to grind. As the partially crushed pieces travel forward, they are further reduced by the smaller balls dropping from a lesser height. Thus the action proportions the energy to the work required.

Improvements in mechanical construction in the cylindrical ball mill in succeeding years included the introduction of babbitted trunnion bearings, the upper half being cored out for the reception of grease lubricant. Cast steel gears and pinions for driving, provision of scoop-type feeders and replaceable metal plates of manganese steel bolted to the interior surface of the shell to resist the wear, were other notable innovations.

Just as the tube mill received its greatest impetus on its introduction to the Witwatersrand goldfields, so the ball mill (in conjunction with the mechanical classifier) made possible the present-day wide application of the flotation process. Without it the close approach to a uniform and controlled size range in the feed necessary to flotation would not have been possible.

Open and Closed Circuit Grinding. With the introduction of the tumbling mill it was customary for many years to grind in open circuit, the ore being passed through the mill once only, the pulp discharge being the finished product as far as grinding was concerned. Thus when a product of fine size was desired, the feed had to be cut back to the degree necessary to achieve the desired results. This was extravagant of power, and led to high consumption of balls and overgrinding of material. Hence the ability of a mill to handle large tonnages was largely nullified. This was eventually rectified by the installation of appropriate sizing equipment in the circuit, thus permitting the mill to be run at full capacity, the sizing device returning oversize to the mill for regrinding. Screens were first used to hold back the coarse material and return it to the mill until ground sufficiently fine to pass through. For coarse separation, screens were satisfactory, but when fine material was required this method became impractical, for the screens soon became choked. The Caldecott diaphragm cone introduced in 1908 was used for some time, but was eventually found inefficient for finely ground pulps, for the reason that fine sand did not settle rapidly enough to prevent liquid breaking through to the apex. When the underflow from the cone was cut down to the extent necessary to prevent the liquid from breaking through, the tonnage of solids for regrinding was so low as to be uneconomic.

Mechanical Rake Classifier. In 1904 J. V. N. Dorr (1871–1962) developed the mechanical classifier at S. Dakota in U.S.A. and the substitution of this machine for screens and

27

cones in closed circuit grinding proved a big step forward. Not only were the particles ground fine enough immediately removed from the circuit and not overground, but efficiency of separation was improved and a reduction in the power load secured. A further advantage was that the separated coarse material was gravity fed back to the mill, no pumps for elevation of product being necessary.

The classifier consists of a rectangular tank 12–30 ft. long by 4–12 ft. wide inclined at a small angle and in which functions a mechanically operated rake. The pulp discharge from the ball mill flows into the lower end of the classifier and is diluted with water, the coarse particles settling to the bottom of the tank, where the movement of the rake continuously pushes them up the slope of the tank until they fall out at the top end and are returned to the mill for further grinding. The fine unsettled material flows over a weir at the lower end and constitutes the feed to the next stage in the process.

FLOTATION

Flotation may be described as the separation of one or more mineral particles from the worthless gangue in a liquid pulp by means of air bubbles. In practice the ore is ground in water to liberate the valuable minerals from gangue, and then agitated with air. Chemical reagents are added to cause the mineral particles to attach themselves to the air bubbles, which then rise to the surface of the pulp where they are removed as a froth. The valueless gangue material does not attach itself to the bubbles and is discarded as a tailing. The weight or specific gravity of the minerals is not the determining factor, for the minerals floated are usually two or three times as heavy as the gangue.

Today by far the greater number of mineral dressing plants apply the flotation process for the concentration of ores, yet the process as it is now generally applied is of comparatively recent origin.

Development of the process may conveniently be divided into three periods:

> 1860–1900   Bulk flotation by means of oil
> 1900–1920   Bulk flotation by means of air
> 1921–date   Selective flotation.

In 1860 an Englishman, William Haynes, in a British Patent described agitation of finely ground sulphide ores with 10–20% their weight of fats or oils. Agitation of the mixture with water caused the oiled sulphide mineral particles to float free from the gangue material. In 1877 the brothers Bessel of Dresden developed a method of floating graphite by adding an oily substance to a finely ground aqueous pulp of a graphitic ore and generating carbon dioxide in it by the action of acid on a carbonate. This treatment generated bubbles which gathered up the graphite and rose to the surface. Carrie J. Everson, an American, in 1885 developed a process for separating the sulphides in an ore from the gangue. She made use of the selective action of the oil for sulphides discovered by Haynes and in addition described agitation of a type that would introduce air into a pulp. Of these processes commonly known as bulk oil flotation and their subsequent refinements little if any use was made in commercial practice.

F. E. and A. S. Elmore, two Englishmen, in a patent dated 1898 mixed one part of ground ore with six to ten parts of water and oil equivalent to one ton of oil per ton of solids. Mechanical agitation of the pulp was followed by introduction of the mixture into hopper-shaped boxes (Spitzkasten) in which the oil concentrate floated to the surface of the aqueous pulp and overflowed the lip of the box. The aqueous portion of the pulp was drawn off from the bottom spigot. Most of the oil in the concentrate was reclaimed by centrifuging. The oil used was the heavy fraction left after distilling petroleum. The use of sulphuric acid to improve selectivity was patented in 1901. This process was the first of its kind to approach commercial application.

By the turn of the century it became apparent that flotation offered advantages over older methods of concentration. High extraction and high ratio of concentration could be obtained at lower operating cost; further, purer concentrates could be obtained, which led to better smelting practice. Engineers, metallurgists and milling men in all the large mining centres began applying flotation to their concentration problems and literally hundreds of patents flowed out of the patent offices of the world. One of the most significant of the discoveries was that of A. Froment of Italy in 1902, who applied the carbon dioxide process of the Bessels to sulphide flotation. The essence of Froment's discovery was the recognition of gas bubbles as the buoyant media for carrying the oiled sulphide mineral particles to the surface of the pulp.

C. V. Potter, in 1902, at the Broken Hill Lead and Zinc Mines in New South Wales, Australia, also recognized gas bubbles as buoyant media. Ore was heated with a dilute solution of sulphuric acid, the acid attacking the ore liberating bubbles of gas which attached themselves to the particles of sulphide carrying them to the surface. This was the first process to operate on a large scale for over 6 million tons of zinc concentrate assaying up to 42% Zn was produced by this process from waste dumps. G. D. Delprat working at the Proprietary mine substituted a salt cake (Na_2SO_4) solution for sulphuric acid. The process subsequently became known as the Potter–Delpratt method.

F. E. Elmore, in British and American patents issued in 1906, subjected the oiled pulp to a vacuum. An air pump was used to generate a vacuum which caused the dissolved air in the pulp to rise in the form of bubbles heavily laden with mineral. This process had its most noted application at the Zinc Corporation plant at Broken Hill, Australia, and in Wales, Cornwall and elsewhere.

It will be observed that the buoyant effect of oils in flotation was being supplemented by the introduction of gas into the ore pulp by reaction of acids or carbonates, by heating or by subjecting the pulp to vacuum. H. L. Sulman, H. F. Picard and J. Ballot in Australia observed that a froth formed when greatly reduced quantities of oil were employed as compared with quantities normal to the process. In British and U.S. patents of 1904–6 they stated that oils in a quantity not exceeding 1% of the weight of ore treated was necessary for the proper functioning of the process and demonstrated that greater quantities of oil were prejudicial to its proper operation. An English company—Minerals Separation Ltd.—was formed by Sulman, Picard and Ballot to exploit this and other processes.

About the same time a flotation apparatus was developed which featured air bubbles generated within a mineral pulp by means of a rapidly revolving impeller. The combination of the use of limited quantities of oil and air bubble production by mechanical means firmly established the froth flotation process, and bubbles ever since have remained the bearer of the vast tonnages of mineral particles.

Selective Flotation. Had advancement in the art of flotation stopped with the advent of bulk flotation, the use and scope of the process would have remained limited. Gravity separation processes in use prior to 1910 did not successfully recover the whole of galena (PbS) from the numerous deposits of Pb/Zn ores and hence there were in existence considerable quantities of galena in the dumps from gravity plants in many parts of the world. Though the flotation process of the time was successfully recovering zinc, much of the galena also floated, the mixed Zn/Pb concentrates being a decided nuisance in smelting. Bulk flotation was incapable of coping with these types of ores and hence exhaustive efforts were made towards a means of effecting the separation of the various minerals.

Sheridan and Griswold, working at the Timber Butte Mill east of the city of Butte, U.S.A., were concentrating on research dealing with the depression and activation of the zinc mineral sphalerite. The ore being treated was a complex one, principally zinc but with values in lead, silver and pyrite. In 1922 a U.S. patent was awarded to these two workers for the separation of lead sulphide from zinc sulphide by the depression of the sphalerite by the use of a soluble cyanide in an alkaline circuit. This was a major breakthrough, for coupled with earlier discoveries that (*a*) pyrite can be depressed with lime and (*b*) copper ion was an activator for sphalerite, the method made available a large number of lead–zinc–iron deposits which formerly were either unworkable or at best were being inefficiently treated by complex gravity flowsheets.

Incidentally bronze impellers used in the early flotation machines were responsible for the accidental discovery of the use of copper sulphide as an activator of sphalerite.* This occurred in 1915 at the old Butte and Superior Copper mine at Butte, Montana, who were treating a lead-zinc ore. There were two banks of cells, one bank being equipped with bronze impellers, the other with iron. It was noted from day to day that the concentrates from the bronze impeller cells ran higher in zinc than the concentrates from the iron impeller cells. The

---

* 'Fifty Years of Flotation', page 86, *Engineering and Mining Journal*, New York, December 1961.

laboratory cell was also equipped with a bronze impeller, and replacement with an iron impeller gave results that were in line with the iron-equipped mill cells. Copper sulphate was now added to the iron-equipped laboratory cell with the result that the concentrate now carried more zinc than either bank of plant cells. Hence the discovery of copper ion as an activator of sphalerite.*

FLOTATION MACHINES

The machine in which the flotation operation is effected consists of an open top vessel through which the pulp is conducted and includes means for simultaneously agitating and aerating the pulp. Depending on the manner by which aeration is effected, two main types of cell can be recognized, viz. the pneumatic cell and the sub-aeration cell.

Pneumatic Cells. In this type of cell air is introduced into the pump through a porous or other medium at the bottom of the cell. The early pneumatic cells such as the Callow cell, introduced by J. M. Callow in 1915, consisted of an open trough 2–3 ft. wide and 2 ft. deep and about 9 ft. long constructed with a false bottom of canvas stretched over a wooden grating.

Low-pressure air from a blower was introduced beneath the canvas, the compressed air as it rose in the form of bubbles, collecting the mineral and creating a froth which overflowed into a launder, the barren gangue discharging over an adjustable weir at the end of the trough. Cells of this type were subject to clogging due to the deposition of carbonates on the canvas fibres from action of carbon dioxide of the air and the subsequent cementation thereon of fine ore. To relieve such clogging the cell had to be emptied at frequent intervals and the canvas brushed with acid and scrubbed with a wire brush. In the MacIntosh cell, introduced in the early twenties, the clogging was overcome by employing a rotating canvas-covered drum mounted near the bottom of the trough. The rotation reduced the tendency to blind, and aeration was more efficient.

In the Forrester cell introduced in 1925 the porous media, and hence clogging, were entirely eliminated, the principle of

---

* It has also been stated that L. Bradford working at Broken Hill discovered in 1913 that copper sulphate greatly increased the floatability of sphalerite.

the air lift being utilized to aerate and circulate the pulp. Air is introduced into the centre of a V-shaped cell by means of a series of vertical pipes spaced at about 6-in. intervals along the length of the cell and supplied with air from a horizontal header. The incoming air mixes with the pulp and owing to the difference in hydrostatic pressure lifts the pulp between baffles. The air is thereby broken up into small air bubbles which with their accompanying mineral load rise to the surface of the pulp in the main body of the cell, the froth overflowing from the sides of the cell, the tailing discharging at the end of the cell over an adjustable weir. The Forrester cell was the first of the air lift machines to find commercial application and for some years enjoyed a considerable vogue.

Sub-aeration Machines. In this type of cell (Fig. 2) a vertical shaft drives an iron impeller at the bottom of the cell. A difference in the manner in which the air is introduced to the cell divides machines into two classes. In one class, the impeller serves as a pump, drawing in air either through a stand pipe surrounding the impeller shaft or else through the shaft itself which has been hollowed out for this purpose. The air is thus mixed with the pulp as it passes through the impeller zone, diffuses into the main body of pulp and rises to the surface when it is removed. In another class, air from an external source is blown into the agitation zone usually directly under the impeller, being distributed through the pulp as before by the rotation of the impeller. The sub-aeration type machine has almost completely displaced the pneumatic cell over the past twenty years or so.

REAGENTS

Flotation requires the use of chemicals which are added to the ore pulp for the following specific purposes: (*a*) Collectors or Promoters. The function of this type of reagent is that of promoting the floatability of minerals in order to effect their separation from the gangue material. They achieve this end by temporarily altering the surface of the minerals to be floated, so as to cause them to adhere to the froth bubbles, by which collection into a froth concentrate is made possible. (*b*) Frothers. The purpose of frothers is to produce a froth which serves as the buoyant and separating medium in which the minerals collected by the promoters are separated from the pulp. (c) Modifiers. Reagents of this class change the surface of

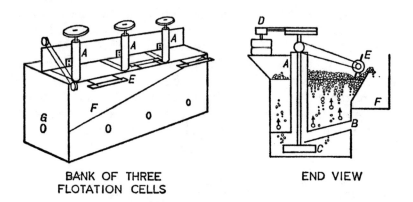

BANK OF THREE FLOTATION CELLS

END VIEW

Fig. 2

Flotation Machine showing (A) Air Pipe, (B) Feed Pipe, (C) Impeller, (D) Motor, (E) Froth Paddle, (F) Froth Launder, (G) Tailing Discharge

minerals so as to modify their amenability to flotation with a particular collector-frother combination.

Collectors. Oils such as petroleum, wood and coal tars were used as collectors up to the early twenties when they were displaced by organic chemical compounds which combined a much more effective collecting power with selectivity. C. L. Perkins in 1921 demonstrated in a U.S. patent that certain definite chemicals, nonoleaginous in nature and containing sulphur or nitrogen (e.g. naphthylamines, toluidines and thiocarbanilide) were effective in collecting mineral particles. This was followed in 1924 by an important discovery made in America by C. H. Keller of the efficiency of the water-soluble xanthates as collectors of sulphide minerals; this marked another milestone in the history of flotation, for by their use it was possible for the first time to effect independent control of collection and frothing, so that flotation changed from a mechanical to a chemical process. Further, as the xanthates are water-soluble, a much more precise control of the dosage was possible, for the small quantities needed are more easily added in dilute water solution than in oil solution.

Xanthates were originally discovered by Zeise in the year 1822 and as the compounds are yellow in colour, they were named after the Greek word 'xanthos', meaning yellow. From the time of their discovery until the beginning of the century xanthates had no commercial use. Shortly after 1900 the rubber industry adopted xanthates in connection with the curing and vulcanization of rubber and later on the compounds found an application in the manufacture of artificial silks. Xanthates are dithiocarbonates, derived from monomolecular quantities of an alcohol in alkali and carbon disulphide.

$$C_2H_5OH + NaOH + CS_2 = C_2H_5ONaCS_2 + H_2O$$

By the employment of different alcohols, i.e. ethyl, isopropyl, butyl, amyl, hexyl, etc., many different xanthates are possible, but in flotation only those containing from two to five carbon atoms are used.

In 1926 another set of soluble collectors was added to the list of flotation reagents. F. T. Whitworth working at the Utah Copper Co. introduced the soluble dithiophosphates, reaction products of phosphorus pentasulphide with various organic compounds such as phenols and alcohols. Following Perkins' and Keller's discoveries the employment of acid pulps rapidly gave way to alkaline circuits, very few acid circuits being in use today.

N. C. Christensen in a U.S. patent of 1923 showed that differential flotation of ore gangue was possible by the use of fatty acids. This group of collectors comprises the acids and salts in which the active radicle is the carboxyl (COOH) group. Fatty acid flotation is known as soap flotation and finds general use for the flotation of alkaline earth minerals such as apatite, barite, fluorite, magnesite, etc.

Frothers. Frothers are added for the specific purpose of creating a froth, and function by lowering the surface tension of the water composing the pulp to a point such that the mineral laden air bubbles may reach the surface of the water intact, and thus form a coherent mass of mineral-laden bubbles known as froth.

The reduction in oil requirements following on the discoveries by Sulman, Picard and Ballot in 1904 indicated that there were differences in the frothing properties of different oils. Specific frothing agents such as alcohols, ketones, and esters were experimented with, and indications were found that frothing properties were associated with definite chemical groups such as OH, CO, COOH, etc.

About 1912 I. W. Wark in Australia discovered the frothing properties of eucalyptus oil (a hydro-aromatic alcohol), and in the U.S.A. pine oil and wood creosote (a phenol) came into use shortly after. In 1920 cresylic acid ($C_6H_4CH_3OH$) was introduced and has become one of the most widely used of all frothers.

The active frothing constituents of these reagents are mono-hydroxylated compounds such as alcohols or alkyl phenols and hence the main polar group utilized in practice is the hydroxyl group (OH).

Modifiers. These reagents modify or control the normal behaviour of minerals in a flotation operation and may be divided into three general classes: (1) Regulating agents. (2) Activating agents. (3) Depressing Agents.

Regulating Agents. In 1923–24, hydrogen ion (pH) concentration as a modifying and controlling factor in the use of collectors came into the picture, it being found that in any particular flotation operation there is in general a pH range in which optimum results are obtainable. The reagents commonly used for pH adjustments are lime and soda ash to increase the pH, and sulphuric acid to decrease the pH. Caustic soda is also occasionally used for pH regulation. Most if not all flotation operations operate best at some pH level and many now employ automatic pH control systems.

Activating Agents. Reagents of this group are used to effect the flotation of certain minerals that are normally difficult to float with collectors alone. The outstanding example of this type of reagent is copper sulphate, used to activate sphalerite and marmatite as well as pyrite and arsenopyrite. Other examples include the use of sodium sulphide to activate lead oxide, carbonate and copper carbonate minerals.

Depressing Agents. The Sheridan and Griswold discovery, in 1922, of the selective depressing action of cyanide was an important advance in the art of flotation; hitherto it had been practically impossible to separate minerals which had similar floatabilities.

The most widely used depressant is sodium cyanide employed in the separation of galena from sphalerite, copper sulphides from galena and pyrite and nickel sulphides from copper sulphides. Other commonly used depressants include lime for depressing pyrite, zinc sulphate as an adjunct to cyanide, chromates for galena depression, quebracho and tannic acid for depression of calcite and dolomite in the flotation of fluorite.

The quantities of the various reagents used in flotation are very small. They are usually expressed in pounds per ton of ore treated, the following figures being typical:

Frothing Agents	0·02–0·10 lb. per ton	
Collectors (Xanthate)	0·05–0·20	,,
Collectors (Fatty Acid)	1·0	,,
Depressants	0·02–0·1	,,
Activators	0·5 –1·0	,,

General Remarks. Flotation was first used on the tailings from gravity separation mills. These tailings generally contained high values, for jigs and tables of the early days were not outstandingly efficient in the recovery of minerals, especially zinc.

In addition they were in a fine state of subdivision, and the tailings constituted a 'natural' for the flotation process. Gravity mills, however, continued to flourish, for smelters were not yet equipped to handle the excessively fine flotation concentrates, and in fact smelters paid a premium for coarse concentrates. Further, the treatment of complex ores by flotation in the early days was unreliable, for the bulk flotation processes provided only a combined concentrate whereas jigs and tables were able to effect some degree of separation.

At first because flotation was a process with unlimited possibilities much secrecy was prevalent amongst the users. This was amplified when in 1914 the Mineral Separation Co. (who owned the Sulman, Picard and Ballot patents) brought suits against the Butte and Superior and Miami Copper Co. for infringement of patents. Both suits were decided in favour of the plaintiff which established their right to collect royalties for the life of the patents.

Litigation in flotation was in fact rife, for it raged for some 15–20 years after its introduction. The *New York Engineering and Mining Journal* in April 1911 said: 'The litigation over the flotation process is a good Kilkenny fight. Everyone who has got a shillelagh is hitting every head in sight. The Elmore people are suing Mineral Separation and the latter are suing back. In fact, almost everybody is suing almost everybody else, whilst concerns that own no patents are being drawn into the vortex.' For this reason operators were hesitant to publish details of their use of and improvements in flotation, and the effect on immediate progress was wholly detrimental. At the same time it was forced on the industry that unsuccessful litigation could be considerably more expensive than successful research. As a result operators working both in the laboratories and on concentrations gradually carried the process forward towards present-day practice.

Concurrently with the advance in flotation, chemistry and metallurgy, mechanical improvements in flotation equipment were also noteworthy, until practically all minerals can now be floated.*

Flotation in Practice. In present-day practice (Fig. 3), the ground pulp is first conditioned with the requisite reagents and then fed into the feed compartment of the first or second cell, being directed on to the rotating impeller where it unites with air. During the agitation, bubbles collect the mineral values, the mineral-laden froth rising to the surface where they are carried forward by the crowding action of succeeding bubbles, removal of the froth being secured by rotating paddles which sweep the froth into launders for conveyance to de-watering filters. The gangue together with any mineral which has escaped flotation in the first cells passes over a weir partition and drops into cell No. 2 so that any unfloated mineral is subjected to the same action and so on in cells No. 3, 4, etc.

Thus the pulp travelling from the feed cell to the discharge of the final cell is continually being impoverished of its mineral content, until finally the froth produced in the last cell is practically barren. The mineral-laden froth containing some 50% water is thickened and then de-watered on a filter down to 5–10% moisture.

* The tin mineral cassiterite is, however, a notable exception.

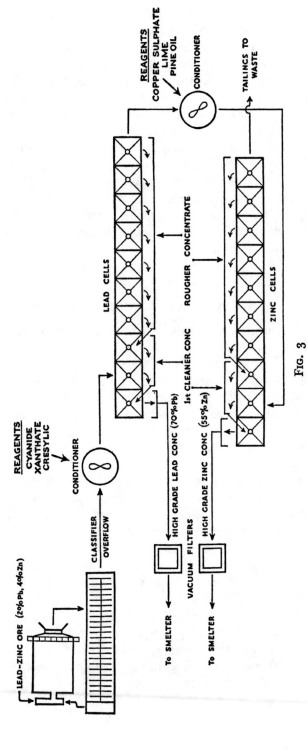

Fig. 3

FLOWSHEET ILLUSTRATING SELECTIVE FLOTATION OF LEAD-ZINC ORE

Circuits. It will be apparent that as several individual cells comprise the flotation unit, many variations in the collection of froth and in the flow of pulp through the cells are possible. The type of circuit is actually dependent on the quantity of tonnage to be treated and also on the grade of mineral, but all circuits have this in common, that the primary or rough concentrate which is first produced is almost invariably subjected to a second or cleaning flotation, thus producing a higher grade concentrate. Many flotation plants also make an intermediate product or middling generally of particles composed of more than one mineral in which the primary grinding has failed to liberate the minerals from the gangue or from one another. In this case the middling is often collected separately and sent back to the grinding circuit for regrinding before being again introduced to the flotation circuit.

GRAVITY CONCENTRATION

One of the simplest operations in mineral dressing consists in subjecting an aggregate of crushed ore to a stream of water, whereby the lighter gangue materials are carried away leaving behind the heavier minerals. This method of mineral separation is made possible by the high specific gravity which is one of the outstanding characteristics of most minerals. In common with lustre this would appear to be the property that first attracted the attention of early man and led to the recognition of metallic ores as such.

The earliest treatises on mining show appliances making use of the principle of gravity separation, and these ancient devices have in many cases, in improved and modified form, continued in use down to the present day.

All wet gravity methods involve the suspension of the ore particles in a fluid medium, separation being effected in the majority of cases by a combination of hindered settling and flowing stream selection. The processes employed to effect gravitational separation are sluicing, strakes, tabling, jigging, and sink-and-float. It is a prerequisite for all forms of gravity concentration that the material to be treated is crushed fine enough for liberation of mineral, and in most cases close sizing is also essential for efficient recovery.

Sluices. In a flowing stream any heavy particles tend to settle and collect on the bed, this action being accentuated by any obstacle which may be present on the bed of the stream. This is the principle of the Long Tom, Rocker and other forms of riffled sluices formerly much used in the recovery of gold from auriferous gravels (page 267) and to some extent still used in the recovery of the tin mineral cassiterite.

Essentially such appliances consist of an inclined trough or launder into which the feed is washed by a stream of water, the gold and other heavy minerals being caught against strips of wood (riffles) nailed at intervals across the bed of the trough, the lighter material being washed away by the stream of water. The trough may be stationary or mounted on rockers, moved in such a manner as to aid the progress of the material along the bed of the trough.

Strakes. One method used in the concentration of gold in early times according to legend was to line the sluices with sheepskins (the Golden Fleece of the Argonauts) any fine gold being caught by entanglement in the hairs. This primitive device in modified form is still used by covering a sloping table with blankets or corduroy, mineral pulp being directed over the table at right angles to the pile, the heavy minerals being caught and retained by the nap. The cloths are removed from time to time and washed to remove the concentrate that has been caught. The device is known as a strake and because of its cheapness was at one time in extensive use for the saving of fine gold; but with the advent of cyanidation its use has declined.

Tables. In the above devices the separatory bed is normally stationary, but a much more effective concentration is attained by imparting motion to the separating surface, for by this means stratification according to the specific gravity of the mineral occurs, allowing of the separate collection of each mineral species. The Rittinger table, introduced about 1850 by the German scientist Bergrath von Rittinger was the first of this class. It consisted of a wooden platform 8–12 ft. long by 3–6 ft. wide, sloped at 6° in the direction of the length and suspended from overhanging supports on four rods at the corners. Motion was imparted in the form of blows or bumps delivered against one side of the table by a cam which pushed the table towards one side, a spring forcing it back until stopped against a bumping post. This action caused the ore particles which had been fed in at the top end and had settled out of the pulp to travel across the table by a series of jerks, while wash water fed in at the head of the table caused them at the same time to move down the slope. The combination of these two actions caused the mineral particles to spread out like a fan in which the different minerals were sorted according to their specific gravities, the heaviest particles being propelled farthest across the table—passing off at the end opposite the driving gear—the intermediate weight mineral next; the lightest gangue materials were the least affected, being carried down by the stream of water and discharging over the edge of the table opposite the feed end. The table received 150–200 bumps per minute, the magnitude of the movement being 1–1½ in. according to the character of the feed. The capacity varied from 2 tons per day for coarse material down to 1 ton or less per day for slime.

The Rittinger table constituted a great advance in concentrating methods and was used in Europe and elsewhere until the advent of the Wilfley table patented in 1895 by Arthur R. Wilfley of Denver, Colorado. The Wilfley was in many respects similar to the Rittinger, but the essential improvement was the provision of wooden riffles, spaced about 1 in. in a direction at right angles to the main flow of pulp and terminating along a diagonal line extending from the forward end of the feed box to the opposite corner. As a result the solids in the feed settled in the troughs formed between the riffles, the heavy particles gravitating to the bottom and being guided by the riffles to the smooth portion of the tables for further washing, any entangled light mineral or gangue being washed down the slope. The shaking motion maintains the particles in a semifluid bed; this, combined with the wash water flowing down the slope, enabled a much better stratification to take place according to specific gravity and size, and hence a much more efficient separation. A Wilfley table will treat 5 tons of fine material or 20–30 tons of coarse material per day using 10–30 gals. per minute of water. The drawback of small capacity in relation to floor space occupied by the shaking table has been to some extent rectified in recent years by the provision of multi-deck tables.

Jigs. If a hand-sieve is filled with crushed ore coarse enough to be retained on the mesh and caused to move up and down in water, it will be observed that after a few movements three layers of material are produced, a lower layer of heavy mineral, an upper layer of the specifically lighter gangue and a middle layer composed of intermediate-gravity and locked mineral particles. This can be regarded as the original form of jigging machine and was in use for the concentration of minerals as far back as the 15th century, as shown by the illustrations in the well-known work of Agricola (1556), *De Re Metallica*.

The vertical pulsating currents of water which are essential for promoting the concentration may be produced in two ways, either as above by moving a sieve up and down in a tank of water, or by having a fixing screen, the water being caused to rise and fall by the reciprocating action of a piston or plunger in an adjacent compartment connected with the space below the screen. The former was the oldest method and continued in use for centuries, the latter or fixed-sieve jig being introduced into Cornwall in 1830 by Petherick* and now constitutes by far the most generally adopted form of machine. Commonly the machine consists of a hopper-shaped wooden box divided into two compartments one of which contains a screen and the other a plunger actuated through an eccentric. Sized ore and water are fed continuously on to the screen, the ore being brought into suspension by the action of the plunger, which forces water up through the screen 150 to 250 times a minute. Under the influence of the impulses the heavy minerals sink on to the screen, the lighter gangue separating out at the top layer where it is swept away by a flow of feed water across the compartment. The size of the screen mesh determines whether or not the mineral particles will remain on the screen or fall through into the hopper. If the size of the screen mesh is made larger than the particles then the operation is termed 'jigging through the screen', the mineral concentrate being drawn off in the hutch below the screen, otherwise the concentrate accumulates on the screen and passes out of the compartment through an opening where it is collected.

Several jigs are usually built into one unit so that the tailing from one compartment passes to the next and is subjected to further jigging action by means of which some of the remaining mineral is removed, the tailing passing into the third compartment, and so on. Three to four jigs commonly comprise one

* *Trans. Inst. C.E.*, Vol. XXX, 1869–70, page 125.

unit. On the sieve it is usual practice to place a coarse bed of iron-shot, the function of which is to keep the gangue out of the concentrate when passing into the hutch. The depth of this bed is generally 1 to 2 in. Needless to say, this bed is of larger size than the apertures of the sieve. Jigs do not usually produce in one operation both a finished concentrate and tailing. They are ordinarily used either to reject rock, i.e. a tailing, and produce a middling for further treatment, or alternatively to make a middling and finished concentrate.

There are many modifications of this type of machine and although they previously had a wide range of usefulness, competition by the flotation process and the sink-and-float operation has caused a decline in their application.

Sink-and-float. If a sample of crushed ore is introduced into a liquid medium whose specific gravity is intermediate between that of the gangue and that of the mineral, the gangue particles will float, the mineral particles sinking. This is the basis on which the sink-and-float method is established, and from which it derives its name. Differential separation of minerals by this means dates back to 1891 when patents were obtained in the U.S. by J. S. Lurie, who suggested the use of heavy liquids such as tetrabromethane, cymene, etc., to effect the separation. The difficulty, however, of obtaining such fluids, and their cost, precluded their industrial application. Patents were taken out later by Conklin and Chance, and by De Vooys and Tromp in Europe. As a substitute for heavy liquids a suspension of finely ground materials in water was specified. The process patented by T. M. Chance in 1917 was the first recognized sink-and-float separation worked on a commercial scale. The heavy liquid employed was a suspension of fine sand in water maintained at a density of 1·5–1·7 by vigorous stirring in a conical tank supplemented by rising currents of water. It was used for the cleaning of coal, the coal floating over the periphery of the tank, the slate, refuse and dirt sinking and being removed from the bottom by a bucket elevator.

De Vooys employed a suspension of barytes and clay, the Barvoys system, as it is known, being used in Europe for coal cleaning.

Such solutions yielding gravities suitable for the separation of coal could not be used for the separation of minerals, for the specific gravity of the suspension used was very much lower than that required for separation of mineral from gangue. The Chance and Barvoys suspensions are limited to an upper specific gravity of $1 \cdot 7$, whereas the specific gravity of ore gangue material is at least $2 \cdot 5$; hence the specific gravity of the suspension must be higher than this for efficient separation of mineral from gangue. It was perhaps natural that attention was directed to the employment of mineral suspensions as a separating medium. In 1934 experiments were begun both in this country and elsewhere in the use of galena suspensions. This mineral has a specific gravity of $7 \cdot 5$ and a workable suspension of $3 \cdot 3$ in water is readily obtained with the finely ground material. In 1936 the American Zinc Lead and Smelting Co., employing a galena medium at their Mascot mill in Tennessee, placed in operation the first commercial sink-and-float plant for the treatment of metallic ores. The equipment (which replaced jigs for primary concentration) consisted of a chamber containing the medium, means for removal of sink-and-float products, screens for reclamation of entrained medium from the products and equipment for reconditioning the reclaimed galena. The washed and sized $(- 4$ in. $+$ 10 M$)$ zinc ore is fed to a cone containing the medium which is maintained at a gravity of $2 \cdot 75$. The gangue material floats and is continuously removed by overflowing a weir. The heavy mineral sinks and is removed by a central airlift which elevates it together with some of the medium. The float and sink then go to screens, the greater part of the adhering medium draining through and being returned to the cone by a pump. The materials then pass to further screens for water washing for removal of any adhering medium. Finished float and sink products are discharged from the screens, the mineral passing on to the next dressing operation, the floats being conveyed to the tailings dump.

The undersize from the washing screens consisting of medium, wash water and fines is de-watered in a thickener, the galena being recovered by tabling. The capacity of the initial unit was 3,500 tons of ore per day, the head feed to the cone containing $2 \cdot 5\%$ Zn, the sink product assaying 15% Zn, recovery being 88%. Approximately 60% of the total mill feed was rejected as a tailing.

In the U.K. in 1935 the firm of Huntington Heberlein, employing a galena medium, erected a pilot plant first at

Barking, Essex, and later at the Halkyn Lead and Zinc Mine in North Wales. The method employed is, however, different in several details from that outlined above.

Removal of sink was accomplished not by an airlift, but by a bucket-type elevator, the material being picked up at the base of the cone and elevated to the cleaning and washing screens. Removal of float from the surface of the medium was effected by means of a drag conveyor, while the contaminated medium was cleaned by flotation, thickened and returned to the cone.

The medium used at the Mascot mill of the American Zinc Co. was changed in 1948 from galena to ferrosilicon. The change resulted in lower tailings assay, higher mineral recovery and lower operating costs. Ferrosilicon, which is an alloy of iron and silicon with a specific gravity of 7·0 had been in use since 1939 as the separating medium for concentrating iron ores. In a very fine state of subdivision it is mixed with water to give the predetermined density. As the alloy is magnetic it is reclaimed from the wash water and fines by magnetic separation. Preconcentration using ferrosilicon as the medium is in general termed Heavy-Media Separation (HMS), the American Cyanamid Co. being the technical and sales representative throughout the world.

The process is widely applied as a preconcentration process on lead and zinc ores, preceding flotation, rejecting up to 60% of the mill feed at a coarse size. Finished products can also be produced as is the case with iron ores, coal and gravel.

ELECTRICAL CONCENTRATION

A further property of minerals that has been utilized to effect concentration and/or separation is their differential response to electrical force. The application to mineral concentration takes one of two forms. Magnetic concentration, by far the most widely applied, takes advantage of the difference in magnetic permeability of minerals; electrostatic separation, which is a much more recent innovation, depends on the readiness or otherwise of a mineral to accept an electrical charge.

Magnetic Concentration. Faraday enunciated the principle that different bodies have different capacities for magnetic induction, and that this induced magnetism may be observed when a body is placed in a magnetic field. Whether a substance responds or not depends on its permeability, i.e. on the ability of magnetic lines of force to pass through the substance. The extent to which this takes place can be seen by reference to the following table, the attractive force of various minerals being expressed on a relative basis with iron taken as 100.

TABLE

RELATIVE MAGNETIC ATTRACTIVE FORCE OF MINERALS

Mineral	Attractive force (Iron = 100)
Magnetite	40·2
Franklinite	35·4
Ilmenite	24·7
Pyrrhotite	6·7
Hematite	1·3
Zircon	1·0
Garnet	0·4
Rutile	0·4
Pyrite	0·2
Chalcopyrite	0·14
Galena	0·04

The table shows that the great majority of minerals are very feebly magnetic, only magnetite, franklinite and ilmenite exhibiting the phenomenon to any large extent.

In practice the crushed ore (dry or wet) is brought into the vicinity of a magnetic field, and the magnetic particles are attracted to the poles, but—for convenience in collection and removal—are prevented from coming into contact with the electro-magnet by a drum or belt of non-magnetic material which encloses the magnets. The magnetic material adheres to the rotating drum or moving belt as long as within the magnetic field; but when removed from its influence by the continued movement of the belt or drum, it drops into a suitable receptacle. The non-magnetic portion of the ore is not attracted and hence follows a different path. Magnetite is the most magnetic of all natural minerals, and as might be expected this material was the first to be concentrated by magnetic means. As deposits of this mineral occur extensively in Sweden and the U.S.A. the first magnetic separating machines originated in these two countries.

One of the earliest machines to achieve practical success was the Swedish Wenström built in 1883. It consists of a horizontal drum built up of alternate lamellae of wood and iron, the drum being rotated about an eccentrically placed electro-magnet located within the drum and covering about half the interior face of the drum. Crushed ore is fed on to the upper surface of the drum and is carried round with it as it rotates; on entering the magnetic field the non-magnetic material is thrown clear and drops off, the magnetic material, however, being held firmly to the surface of the drum until it is carried past the magnetic field when the magnetics are released and fall off. The drum rotates at about 30 r.p.m. and requires 15 amps at 110 volts, the capacity of a 30 in. diameter by 24 in. width drum being about 5–10 tons per hour. A refinement in operation was the development a few years later of a drum separator which magnetically agitated the material, disengaging the non-magnetics and thus allowing a cleaner concentrate to be attained.

C. M. Ball and D. Norton in America enclosed a sinusoidal stationary magnet within a rotating cylinder of brass, the magnets being so wound that adjacent poles were of opposite polarity. Electrical connection was made through the drum shaft, which was hollowed out. When sized ore is fed on to the rotating drum, each particle of ore in the magnetic portion of the feed tends to present its proper pole to each succeeding pole of the magnets, and in so doing is turned end to end, thus disentangling itself from the non-magnetic, the result being a winnowing action in which the gangue is released and falls off. This produced a much cleaner concentrate and by employing two drums in series, the second drum retreating the rough concentrate from the first, a very high grade concentrate was produced.

By the employment of a field of high magnetic intensity, J. P. Wetherill of New Jersey, U.S.A., made an important advance in 1896 when he showed that minerals of low magnetic susceptibility could also be separated by magnetic means. Wetherill obtained a magnetic field of the requisite intensity by the employment of wedge-shaped collecting pole pieces, thereby increasing considerably the concentration of the lines of force in the narrow edges of the pole piece. By bringing the pole pieces very close together the field intensity was increased still further, the ore particles being brought very near to the collecting pole. The chief features of the machine consist of two or more horseshoe-shaped electro-magnets; the two magnets of a pair are wound so that poles of opposite polarity face one another, and are so adjusted that the air gap between the poles can be brought very low, induction thereby being of a very high order. When the machine has two pairs of magnets the second pair are wound to obtain a more intense field by increasing the turns of wire, thus in a machine having six poles, the first pair of magnets might have 30,000 ampere turns, the second 60,000, and the third 100,000, each pair of poles being provided with a separate ammeter and rheostat for purposes of current regulation.

Passing between the poles of the magnets is a conveyor belt carrying the material to be treated. Cross belts at right angles to this main belt also pass between the poles. As the finely ground ore passes between the poles of the first electro-magnet the magnetic particles are attracted towards the upper poles against the force of gravity, and jump towards it. They are intercepted by the underside of the cross belt which removes them from the vicinity of the magnets and drops them into a bin. The non-magnetics pass on and are discharged over the head pulley of the feed conveyor belt.

The machine operates only on fine ore which must be bone dry in order to allow free movement between the ore particles. Capacity of the machine varies between $0 \cdot 1$ ton per hour for weak magnetics and 5 tons per hour on more strongly magnetic material. The machine is in use for separating wolfram from cassiterite, ilmenite from magnetite, franklinite from zincite, etc.

WET MAGNETIC SEPARATION

Wet magnetic separation where it can be applied possesses several advantages. Fines are more easily separated and better products are obtained because water causes a superior dispersion of the particles. Further, in dry separation dust tends to cling to the magnets, thus interfering with efficiency. Wet separations are, however, limited to a maximum size feed of about $\frac{1}{8}$ in.

The first of these wet magnetic concentrators employed a revolving drum similar to that used in the Wenström machine. The Grondal drum wet magnetic separator devised by Dr. Grondal of Stockholm in 1902 consisted of a sinusoidal electro-magnet occupying about 120° of the circumference of the drum in which it is mounted. The pole pieces of the magnet are directed downwards with the object of creating a powerful field at the lowest point of the drum. The surface of the drum is composed of alternating strips of brass and iron. The drum rotates in an inverted pyramid-shaped hopper with its lower face just immersed in water. Feed is led into the box and prevented from settling by a current of water entering from below. The magnetics are attracted by the rotating drum, lifted clear and released outside the hopper where the magnets terminate, the non-magnetics do not adhere and are drawn off through suitably arranged ports. The modern form of this machine, extensively used in Scandinavia and the U.S.A., consists of a brass cylinder 2–3 ft. in diameter and 6 ft. in length, the lower face of which as before is submerged in water. Magnets may be either of a permanent type or electro-magnetic supplied with currents from an outside source by means of a cable through the drum shaft hollowed out for the leads. Several advantages accrue from using the permanent type of magnet. The initial cost is lower because transformers, rectifiers, wiring, etc., which make up about one-third of the cost of the electro-magnetic are eliminated. Maintenance is reduced, running time is increased and the drums are more watertight. The drums run at a speed of about 30 r.p.m. with capacities in the order of 4 ton/ft. of drum width/hr.

Belt type wet separator. In this type of machine the attracted material is kept separated from the magnets by a belt in a manner somewhat similar to that in the Wetherill machine.

The most widely used wet type belt machine at the present time is the American Dings-Crockett separator, this being a development of an earlier machine, the Dings-Roche, the latest development being due to R. E. Crockett. The Dings-Roche combines wet concentration with magnetic agitation based on 4 stationary magnets with alternate poles suspended below an endless belt. The belt was set at a slope of about 30° the feed being introduced on to the belt, when it met a stream of water, magnetic material being carried up the belt being held by the electro-magnets and discharging over the top whilst non-magnetics were washed down over the tail pulley. In the present-day Crockett the feed suspended in water is introduced to the lower portion of a rotary endless belt, this part of the belt being allowed to sag as it passes beneath the surface of water contained in a tank. Multipolar magnets are so arranged that their poles face the curve of the feed portion of the belt. As the ore enters the feedbox, it is washed up against the bottom belt, magnetic attraction holding the magnetic ore while the non-magnetic material falls by gravity into a tailings hopper. The magnetic material continues with the belt as it emerges from the water, passes over a de-watering tank and is finally discharged by water sprays as it moves out of the magnetic field. The coils are sealed in transformer oil in a stainless steel case which is corrosion and abrasion resistant. The separator will handle ore in the size range $\frac{1}{4}$ in. downwards with an efficiency of 99%. Machines with capacities of up to 100 tons of ore per hour are available.

ELECTROSTATIC CONCENTRATION

Electrostatic concentration exploits the difference in electrical conductivity of minerals. Every mineral will conduct or take an electrical charge on its surface if brought into contact with a source charged at a high potential. The readiness with which minerals do this varies widely and upon these differences is founded the art of electrostatic separation.

The first commercial exploitation of this principle was achieved by L. I. Blake and L. N. Morscher of Kansas who marketed an electrostatic separator in 1901. The machine consisted of a series of poles in the form of insulated metal rotating rollers charged to a potential of 20,000 volts and arranged below each other. In the ore feed, minerals of high conductivity assumed a charge similar to that of the highly charged surface of the roller and were repelled, following a trajectory beyond the normal gravitational fall, whilst the poor conductors were virtually unaffected and fell straight down. Adjustable splitters divided the falling particles into conductors, middlings and non-conductors. The process was repeated on the conductors and non-conductors, the final products being

51

conductors, middlings and tailings, the middlings being recirculated. The machine found a field of usefulness in the freeing of zinc blende from iron pyrites (for whose separation gravity methods were useless owing to the specific gravity of the two minerals being nearly equal), zinc blende from galena, and for concentrating graphite and molybdenite.

In the above type of machine the degree of separation was small, being controlled largely by close regulation of splitters coupled with repeated treatment of the products. During the next decade many patents were filed which sought to achieve a wider divergence in separation between conducting and non-conducting mineral particles. Thus, in America, C. E. Dolbean and others employed a sharp-edged electrode maintained at a high potential adjacent to and parallel with the material conveying metal roller, the mineral particles thereby being subjected to electron bombardment. As a result the minerals of poor conductivity assumed a charge and were 'pinned' to the roller until outside the electrical field, when they dropped off. The conducting particles, on the other hand, did not readily assume a charge and hence did not adhere to the roller, following a normal trajectory.

C. H. Huff* of Brockton, Mass., used round smooth-surfaced electrodes, the material to be separated being brought into the electrostatic field on a grounded conductor (connected to the ground) revolving drum, in front of which, and parallel with it, was positioned a charged electrode. This electrode at first consisted of glass tubing in the centre of which was a $\frac{1}{8}$ in. diameter charged wire. Breakages of the glass tubing were heavy and replacements constituted a welcome source of income until it was discovered that the bare wire alone functioned quite as efficiently.

In 1905 Huff purchased the Blake–Morscher patents and formed the Huff Electrostatic Separation Co. of Boston.

Many plants installed Huff's machine, about 150 alone being in use on zinc dressing plants for removing zinc blende from associated minerals. The advent of flotation in the first quarter of the century greatly curtailed the scope of the electrostatic operation and in 1926 the Huff Co. had to go into liquidation.

In 1935 the Ritter Co. Inc., Rochester, New York, became

* Huff, who has been described as the father of electrostatic separation, was not an engineer, but in the real estate business. He died in 1915.

52

interested in the Huff process and in collaboration with H. B. Johnson after intensive research redesigned the separating equipment which considerably broadened the field of the process.

It was first applied on a large scale in the early 1940s to the Florida beach sands by the Rutile Mining Co., Jacksonville, Florida. These sands contain ilmenite, rutile, zircon and monazite, and although separation of ilmenite from rutile was possible to some extent by magnetic means, the separation of rutile from zircon was found possible only by electrostatic methods. J. H. Carpenter, the engineer at the above company, must be credited with the first development of the high tension technique based on high electrical potential, and the efficient use of ion charging.

In its modern application the electrostatic method effects separation of minerals not readily performed by flotation or other mineral dressing processes. It has achieved prominence in the treatment of titanium-bearing sands, the titanium conductor minerals rutile and ilmenite being separated from the non-conductors zircon, monozite and garnet. It has also found considerable scope in the separation of the tin mineral cassiterite from columbite.

In the modern type of machine, separation is brought about by convective charging in which the ore particles are sprayed with a discharge from a sharp-pointed electrode. Subsequently the particles are subjected to a high intensity electrostatic field, separation taking place according to the sign and/or magnitude of the charge they have acquired. To create the necessary dense high voltage discharge a beam type electrode is used. It consists of a fine wire maintained at a high potential (20,000–40,000 volts), placed adjacent to and parallel with a larger diameter electrode, with which it is in electrical contact. The combination of the large and small diameter electrodes creates a very strong discharge which may be beamed from the wire electrode in a definite direction and concentrated to a very narrow arc, the effect on the minerals which pass through the beam being very strong. Although the ion movement creates a strong conductive field between the wire electrode and the material conveyor, between the latter and the larger electrode there is no ion movement, only a static field created by the difference in voltage.

When ore is fed on to the feed roller it first enters an ionization field where the ions from the wire charge the ore particles. The more conductive minerals discharge the charge to the grounded roller faster than the less conductive minerals and hence take on the polarity of the roller. With the further rotation of the roller the conductors enter the static field and here are drawn away from the roller towards the larger electrode which has the opposite polarity. Hence the conducting particles leave the roller, the trajectory, owing to the electric force, being beyond that of the normal gravitational trajectory. The non-conducting particles on the other hand lose ions very slowly to the grounded roller and hence retain their charge which is opposite to that of the roller, and as a result are held (pinned) to its surface, and are carried by the roller's rotation to a point beyond the action of the field where they fall off. Conducting minerals are thus separated from non-conducting. Adjustable knife edges may be set to divide the falling particles and also provide a middling product for recirculation.

The modern type of machine having 60 in. long rollers has a capacity of about $2\frac{1}{2}$ tons of feed per hour. Bone dry feed is essential not only to ensure a free flow of the material in the electrostatic separation field, but also because moisture can alter the electrical behaviour of minerals, for the electrical conductivity is purely a surface condition.

BIBLIOGRAPHY

1. RICHARDS, R. H., *Ore Dressing*, Vols. I and II (1903); III and IV (1908). McGraw-Hill Book Co., New York.
2. LOUIS, H., *The Dressing of Minerals* (1909). Edward Arnold, London.
3. TAGGERT, A. F., *Handbook of Ore Dressing* (1927). John Wiley & Sons Inc., New York.
4. HOOVER, H. J., 'Concentrating Ores by Flotation', *Mining Magazine*, London, 1916.
5. JOHNSON, H. B., 'History of Electrostatic Separation', *Engineering & Mining Journal*, September, October, December 1938.

CHAPTER THREE

PYROMETALLURGY

PYROMETALLURGY is that branch of extractive metallurgy which deals with chemical reactions at high temperatures commonly attained by the burning of carbonaceous fuel. It is by far the most important of all extractive processes and ranges from a preliminary roasting treatment as employed with lead and zinc ores to the high temperature required in the smelting of iron and copper.

During the past hundred years the pyrometallurgy of most metals has been subject to many changes, and even where the fundamental processes themselves are unchanged, design and construction of machinery and appliances in general have been revolutionized.

ROASTING

Roasting consists in the heating of ores below their fusion point with access of air in order to produce chemical or physical change so that the material will be more amenable to subsequent smelting or other processing. Thus in the case of sulphide ores it is usually necessary to eliminate sulphur, otherwise difficulties arise during the subsequent smelting process.

The blast furnace, the most important unit in iron and lead smelting, necessitates the use of a coarse charge, the necessary agglomeration of fine material being carried out by a special roasting technique. In order to meet the differing requirements many different types of equipment and methods are in use.

Heap Roasting. This is the earliest form of roasting and was in extensive use during the last century for the treatment of copper ore. It was formerly practised at such large plants as Rio Tinto in Spain, Mansfield Copper in Germany, Sudbury in Canada, and the Tyee Smelter, British Columbia.

As the name implies, the method consisted of piling the ore

(sulphide) into heaps, enough wood being incorporated to ignite the sulphur. In making a heap, a bed of about 1 ft. thick of wood is first laid down, and upon it the coarse ore is dumped. This is then covered with fine ore, transverse channels being left in the wood and ore to facilitate combustion of the wood and the oxidation of the sulphides. The size of the heap varied according to the manner of handling the ore, the availability of suitable level ground and the percentage of sulphur in the ore. An average might be 50 ft. × 100 ft. × 10 ft. in height, the contents of which would be about 2,000 tons. The time required for completion of roasting would likewise vary with the size of the heap, the nature of the ore and the amount of sulphur left in the finished material. A small heap might burn for a few weeks, up to three months or so being required for large heaps. At Rio Tinto the roasting had for its object the formation of copper sulphate.

$$2CuS + 7O = CuO + CuSO_4 + SO_2$$

which was then leached out of the heaps *in situ* with water. The metal in the solution was recovered by passage over scrap iron, copper being precipitated as 'cement' copper ($CuSO_4 + Fe = Cu + FeSO_4$). In British Columbia at the Tyee Smelter the ore containing $16 \cdot 6\%$ of S was roasted down to 6% S in 300-ton heaps, the time required being about three weeks.

Stall Roasting. Heap roasting produced large volumes of sulphurous fumes which was fatal to vegetation over wide areas, and led to much litigation. This brought about the use of small bricked enclosures open at the front known as 'stalls'. They were built back to back in rows with a flue running between the rows leading to a chimney stack. Measurements commonly were 8 ft. × 10 ft. × 6 ft. high, capacities being 25–30 tons each. The stalls which were filled (and emptied) by hand were charged as for heap roasting, the ore being laid on kindling wood. The sulphurous fumes passed via the flues to the chimney stack from whence they were discharged high up in the atmosphere. Roasting in stalls was more efficient and the time required considerably reduced. The method was however laborious and obnoxious to operate and eventually gave way to more efficient methods.

Hearth Roasting. The type of furnace in use in the mid-19th century for roasting sulphide ore was of the reverberatory type, consisting of a refractory brick chamber, the hearth being up to 60 ft. long by 16 ft. wide surmounted by a low arched roof. Door openings cut in the side walls were used for rabbling the ore on the hearth. A firebox was located at one end. In some furnaces the hearth was divided into two, three or four sections each having a step of about 2–4 in. on the side towards the firebox the object being to separate the ore during the various phases of the roasting operation. Ore was fed into the furnace on to the hearth farthest from the firebox and moved along by hand by rabbling passing from the coolest to the hottest portion of the furnace, the degree of roasting being regulated by the rate at which the ore was advanced through the furnace so as to avoid fusion or clotting. The capacity of such furnaces was about 10–30 tons of ore, roasting from 33% S down to 6%, at a fuel consumption of about 2 tons of coal. Two men worked a twelve-hour shift. The work was arduous, hot, exhausting and expensive; accordingly every effort was expended in America, where wages were high, to develop a mechanical type of furnace in which the movement of the ore along the hearth was accomplished by rabbles or rakes worked by machinery.

Cylindrical Furnaces. In 1850 A. Parkes (1813–90), a British chemist, patented the first cylindrical multi-hearth roasting furnace which embodied many of the features of the present-day mechanical roasters. It was a brick structure composed of two superimposed brick hearths 12 ft. in diameter over which rotated rabbles, set at an angle and attached to an iron vertical shaft passing up the centre of the furnace and actuated by a bevel gear at the bottom. Ore fed to the outside of the top hearth was raked the full area of the hearth to the centre, where it fell through openings to the second hearth, where the rabbles were set to rake the ore to the periphery, and from thence was discharged through further openings. Although the actual roasting operation was conducted in an efficient manner, and was economical in labour, the brick structure suffered severe stresses from the rabbling action. Having no reinforcement the stresses resulted in a breaking down of the furnace structure through the opening up of numerous joints and cracks which

57

eventually led to the collapse of the furnace. This defect retarded the development of the furnace until 1873, when J. S. MacDougall of Liverpool corrected the structural difficulties by constructing the whole furnace of steel, with the walls and

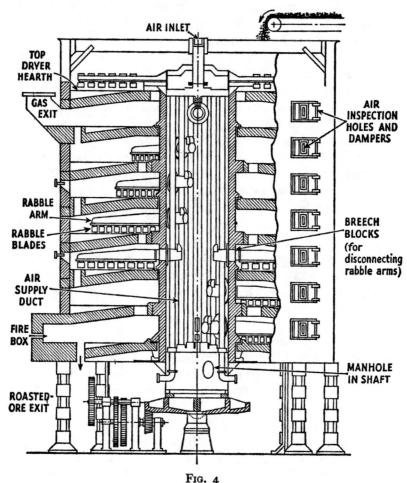

AIR INLET

TOP DRYER HEARTH

GAS EXIT

AIR INSPECTION HOLES AND DAMPERS

RABBLE ARM

RABBLE BLADES

AIR SUPPLY DUCT

FIRE BOX

ROASTED-ORE EXIT

BREECH BLOCKS (for disconnecting rabble arms)

MANHOLE IN SHAFT

FIG. 4

WEDGE TYPE ROASTING FURNACE

hearths lined with brick. The original MacDougall furnace was 18 ft. high by 16 ft. diameter, and had six hearths. The rabbles were of cast iron with internal water cooling. Due to the efficiency of the rabbling action such large volumes of dust were created that the furnace was eventually given up in this country.

In 1892 Evans and Klepetko at the Great Falls Smelter of

the Anaconda Copper Co. brought out an improved Mac-Dougall roaster which became known as the Evans–Klepetko MacDougall furnace. It was mechanically more efficient than the original and roasted 40 tons of concentrate per 24 hours. No fuel was required, the heat of oxidation of the sulphide ore being sufficient to carry the reaction through to completion. J. B. Herreshoff (1834–1930) in America replaced the rigidly attached rabble arms of the earlier design with detachable arms, thus permitting them to be removed through the hearth doors when it was necessary to effect repairs, without closing down the furnace. Water cooling of the rabbles was replaced by air cooling, thus allowing this preheated air to be utilized in the oxidation of the ore.

Wedge (Fig. 4) introduced a wide accessible air cooled central shaft 4–5 ft. in diameter which permitted entry in order to effect replacement of rabble arms and blades. The rabble arms are attached to the shaft by cast iron breech blocks designed to permit quick replacement. The shaft revolves on rollers supported on pedestals, being driven by means of bevel and worm gearing. The furnace itself is usually supported on structural steel columns about 6 ft. high to allow of the discharge of the roasted iron to cars. It is built in several sizes, the capacity of a seven hearth 30 ft. high by 25 ft. diameter furnace roasting pyritic copper ore being in the region of 100 tons per 24 hours.

The chief drawback of the Wedge and Herreshoff furnaces which are today the most widely used type of roaster employed in the metallurgical industry, is the large amount of dust produced due to the rabbling action and the ore falling from hearth to hearth. This gives rise to a tendency to deliver a dirty gas. The Cottrell process of electrostatic dust precipitator has, however, overcome these difficulties to a large extent.

BLAST ROASTING (SINTERING)

The chief disadvantage of the multi-hearth mechanical type of furnace is that the product being in a fine state of subdivision is not in a proper physical condition for charging to blast furnaces. Further, large quantities of flue dust are produced which have to be recirculated if loss of metal values is to be avoided. An answer to these drawbacks was provided by the advent of blast roasting in 1896. T. Huntington and F. Heberlein, while working at a lead smelter in Italy, discovered that if partially roasted hot ore was charged to a suitable vessel and an updraught of air blown through the charge, roasting takes place, the product being an agglomerated porous mass of material, sulphur being reduced to a low figure. The importance of the discovery lay in the fact that (1) a method had been discovered of desulphurizing ore, which was much more efficient than that employed in the mechanically rabbled type of furnace. (2) Feed much more suitable for blast furnaces had been produced. (3) Very little dust was formed. The results were such that the process was widely adopted.

In order to secure satisfactory operation the heat developed by the oxidation must not be excessive, otherwise fusion of the ore particles takes place with the result that free access by the furnace gases is no longer possible, further roasting action thereby being inhibited. In order to control this tendency Huntington and Heberlein preroasted the ore, thereby reducing part of the sulphur radicle, and in addition added lime to the mix to absorb heat.

In its large-scale application the process was carried out in hemispherical cast iron pots provided with a false bottom or grate below which was an opening for admission of air. The pot was mounted on a truck and fitted with trunnions so that it was capable of being inverted for emptying the charge. A wood or coal fire was started on the perforated grate which was then covered with some hot roasted ore. The partially roasted ore mixed with lime was then charged. The pot was covered with a hood for carrying off the voluminous fumes which are formed and the air blast turned on, at first under a small pressure of about 4 oz. which was gradually increased to about 20 ozs. When finished the pot was dumped and the cake broken up by sledging. The time required for blowing varied from 6–12 hours according to the nature of the ore.

Although enjoying a wide vogue at that period the H and H method suffered from the disadvantage that the process was a batch one, and much labour was required in charging and discharging the pot, and breaking up the sinter into lumps of suitable size for the blast furnace. An obvious solution was to develop some continuous method of production and to make the sinter thin enough to be easily broken. Starting with this idea A. S. Dwight (1864–1946) and R. L. Lloyd at the Greene

Consolidated Smelter in Cananea, Brazil, discovered that whereas it was practically impossible to sinter a thin layer of ore by updraught on account of the agitation of the particles in the upper portion of the layer caused by the air currents, it was feasible to sinter the entire layer into a coherent cake by the simple expedient of reversing the draught, drawing the air downwards through the layer of ore on the grate upon which it rested. In addition, by shifting the ignition source from the lower to the upper surface and causing the grate to move, it was possible to change the process from an intermittent to a continuous one.

The first commercial machine started production in 1906 and consisted of a travelling grate in the form of an endless conveyor made up of shallow iron boxes called pallets, all independent of one another, each having wheels running on guide rails. At each end of the machine a pair of large sprocket wheels driven by suitable gearing engage the wheels of the pallets and thus push the train of pallets along. The charge is loaded into the pallets from a hopper to a depth of 6 in. or so, combustion being initiated as the pallets are pushed forward by ignition from an oil or gas burner situated at one end of the track and just above the pallets. Combustion spreads rapidly through the charge by air which is drawn through the charge by a windbox which is located under the upper pallet track and connected to a suction fan. As the sinter passes out of the ignition region, the pallets drop the charge at the far end of the machine and return along the lower track to repeat the cycle.

The first machines were provided with pallets 12 in. wide and an effective hearth area (area of windbox) of some 70 sq. ft. and attained a capacity of 70–100 tons per 24 hours. Today machines are built with pallets 12 ft. in width and 200 ft. long with a capacity of 1,000–2,000 tons per day. Very little labour is required, an operator and helper being able to take care of several machines.

When the material contains no fuel constituents such as sulphur contained in sulphide ores, the deficiency is made up by incorporating carbon in the form of coke fines or powdered coal. This practice has been widely adopted in ferrous metallurgy, the fine material resulting from the screening of iron ore being mixed with coke fines and consolidated into a form dictated by the demands of the blast furnaces.

FLUIDIZED BED ROASTING

When a gas such as air or oil vapour is passed through a bed of finely divided solids at a suitable velocity, the powder dilates, behaving very much as a fluid and can be readily caused to flow from one vessel to another.*

This principle was first applied in the petroleum industry. Trouble being experienced from continual fouling of the catalyst used in cracking by a coating of carbon, mitigation of the nuisance was eventually achieved by a process known as fluidization. In this process, the carbon-coated catalyst was kept in a state of suspension by an upward flowing stream of hot air, the carbon being burnt off, the clean catalyst continuously overflowing the regenerator and being returned to the reactor.

The Dorr Company of America in 1945 adopted the fluidization technique to the roasting of ore. The roasting chamber consists of a large vertical brick-lined steel cylindrical reactor vessel into which air is blown beneath a perforated steel grate at the base. Combustion having been initiated on the grate by a burner, the mineral concentrate is admitted at one side by a screw conveyor. Oxidation takes place, the roasted concentrate dilated by the air continuously overflowing from the bed of the reactor at a point opposite the feed to a hopper outside. Gases pass from the top of the reactor through cyclone dust collectors from which the deposited dust joins the calcine.

The intimate gas/solid contact in the fluid bed reduces the amount of air required for combustion to a minimum; hence the gases are richer in SO_2 than in other types of roaster. Further, it is possible for the feed to be pumped into the reactor as a pulp, thus avoiding the necessity for drying.

The process is now in use for roasting zinc and copper concentrates, for roasting arsenopyrite gold ore before cyanidation, and for production of sulphur dioxide from pyrite for sulphuric acid manufacture.

SMELTING

Smelting may be defined as a process for separation of a metal or metal compound from ore in which chemical change combined with liquefaction occurs in the materials being treated.

---

* When the medium is water the solid will likewise dilate, fluidization of this type occurring during jigging.

Substances known as fluxes are usually added to combine with the gangue material in the ore to produce low melting-point slags. Differences in specific gravity of the molten metal and the fused slag renders possible separation into two layers, so that each may be withdrawn from the furnace at their own level. Smelting does not provide a finished product, for the crude metal obtained generally contains impurities originally present in the ore and refining is necessary.

In order to effect the chemical changes necessary for the liberation of the metal, the ore is subjected to the action of heat at temperatures above the melting point of the metal, the reaction taking place in a refractory structure designed to withstand the high temperatures employed. The type of furnace employed is dictated by the physical and chemical nature of the ore and of the resulting metal; of the many types employed the following are the more common.

Blast Furnace. This type of furnace is distinguished by the fact that the ore and fuel are in intimate contact, the fuel being burned by means of a blast of air (hence the name) passed through the charge. Since the air blast must be forced through the charge, this latter must of necessity be in lump form otherwise blockage of the air blast might occur. Coke is generally employed as the fuel. The blast furnace has always been closely associated with iron ore and lead smelting but it was also at one time prominent in the smelting of such non-ferrous ores as copper and nickel.

Reverberatory Furnace. In general the reverberatory is composed of a shallow hearth with side and end walls surmounted by a low arch roof. It differs from the blast furnace in that the ore and fuel are not in intimate contact, but are kept separate, the products of combustion of the latter passing over the ore contained on the hearth, heat being mainly reverberated or radiated from the roof. Firing may be with pulverised coal, oil or gas, coal as such not being employed in the modern furnace. Having no air blast, a relatively quiet atmosphere prevails inside the furnace chamber permitting successful treatment of finely divided ore. It is extensively used in the smelting of steel, tin, nickel, but it has reached its greatest development with copper. Originally adapted to copper smelting in Wales in the

18th century the small reverberatories 12 ft. or so in length smelting a few tons of ore per day have increased in extent to the present day huge 100-ft. furnaces.

Open-Hearth Reverberatory. Shortly after Bessemer had invented the converter, a further method for making steel was made possible by Siemens' invention of the open-hearth furnace. The name 'open hearth' is derived probably from the fact that while the hearth is covered with a roof, it is accessible via doors in the side walls through which the raw materials are charged. By 1868 open-hearth steelmaking had been successfully developed in Great Britain and elsewhere, until today it is responsible for about 80% of total world steel production.

The furnaces are of the reverberatory type, the hearth consisting of a refractory-lined steel pan, enclosed by front, back and end walls and spanned by a roof which arches over the hearth from the front to the back wall of the furnace. The high temperature (1600° C.) required to keep the charge molten was in the case of the Bessemer process supplied by the oxidation of the impurities (Mn, Si, C) in the pig iron. The attainment of such a temperature in the reverberatory under the conditions prevailing at that time was impracticable, but this was overcome by Siemens' further invention of the gas producer coupled with his application of heat regenerative principle—the heat of the waste gases being utilized for preheating the gasified fuel and the air required for combustion. The regenerators were located under the furnace and consisted of two pairs of enclosed chambers in which firebricks were arranged checkerwork-fashion in order to expose a maximum of heating surface. One of Siemens' early difficulties was that his regenerator system was so efficient that the refractories first employed were unable to withstand the high temperature, and it was not until good quality silica bricks were available that the furnace was a commercial success.

Converter. The converter consists of a refractory-lined vessel in which oxidation of molten metal occurs by air blown through it. The reaction is self-sustaining, no external heat being necessary. It has two important applications (1) Converting of pig iron into steel, (2) conversion of copper matte into blister copper.

Steel Converter. Sir Henry Bessemer in England and William Kelly in America in the early 1850s by their invention of the pneumatic process made possible the first great advance in the manufacture of steel.

The original converter in which the process was carried out consisted of a pear-shaped refractory vessel somewhat similar to a foundry cupola, at first rigidly fixed but later mounted on two trunnions about which it was capable of being rotated for the purpose of charging and emptying. One of the trunnions was hollow to allow for the air blast, which entered the bottom of the converter via a number of openings or tuyères.

Many years of experiment and heavy capital expenditure were to be incurred before success was achieved. Even then it was not universally applicable for the process was not then capable of dealing with phosphoric iron ores; in fact it was not until 1878, the year that Thomas and Gilchrist invented the basic process, that it became generally accepted. Their discovery of a suitable lining for the converter that would permit the removal of phosphorus, made possible the utilization of the extensive deposits of phosphoric ores of the Continent (Lorraine, Luxembourg, Belgium, Saar, etc.).

In this country and America the Bessemer Converter did not make the same progress, for Sir William Siemens in 1867 invented the open-hearth process. This rendered possible the utilization of scrap iron which was then plentiful and cheap, and the steel produced was of higher reliability and of superior quality to Bessemer steel. For these reasons the open-hearth furnace quickly became the predominant steelmaking method.

Copper Converter. The conversion of copper matte to copper was originally carried out in Wales in reverberatory furnaces. The process was slow, the capacity of the furnace limited and the labour demands severe. The reaction was essentially an oxidizing one and Pierre Manhes in France conceived the idea of adapting the Bessemer steel converter to the conversion of matte; in 1884 an agreement was signed by the Anaconda Copper Co. of America with Manhes to exploit the invention.

These early converters were lined by hand with a wet slurry of ground silica and fireclay. Balls of the mixture were pounded into position by a man working inside the converter. Later this manual method was replaced by a mechanized one, the converter shell being carried by an overhead crane from its position in the converter aisle to a relining shop equipped with crushing and mixing facilities for the preparation of the slurry and tamping machines for pounding the slurry into position. The corroded portion was chipped away and replaced by the slurry, which was tamped down in the bottom of the shell to a depth of 6 in. or so below the tuyère line. A wooden template or mould was then positioned inside and the slurry rammed between it and the shell, until a thickness of up to 3 ft. had been built up. (It was desirable to get as thick a lining as possible in order to make it last the maximum length of time.) Mould was removed and the shell subjected to a slow fire in order to dry out the lining. The converter was carried back to the aisle and was then ready for operation. The lining forming an integral part of the process was naturally rapidly consumed, heavy expense being incurred for the constant renewal, one lining lasting only 4 to 6 charges of matte. The method also handicapped reverberatory operation: it was manifestly more economical to treat only high grade matte, for low grade would contain a high proportion of FeS requiring a correspondingly increased amount of silica, thus reducing the life of the lining still further. Many attempts had been made to develop a more durable lining than silica, but without success until W. H. Peirce and E. A. Smith introduced (in 1910) at the Garfield Smelter, Utah, their magnesite brick lining. The lining was of 9 in. magnesite bricks, except at the tuyère belt, where 18 in. were used. The successful evolution of the basic converter was of great importance to the copper smelters for its advantages over the acid-lined type were many: the frequent relining necessary with the acid type converter was abolished; vessels could be made much larger, enabling larger tonnages to be treated; there was no undue restriction as to grade of matte; blister copper was produced in much less time, and copper loss in the slag was lower.

DUST COLLECTION

Roasting and smelting operations form large quantities of dust and fume which pass out of the furnace with the waste gases. As the potential metal loss from this source may be considerable, means have to be provided for the recovery and collection of such material.

Dust as generally understood consists of the finer particles of the components of the charge (ore, flux and fuel) varying in size from ¼ in. down to fine powder. Fume consists chiefly of material such as lead sulphate, lead and zinc oxide, arsenic, etc., which are vaporized in the furnace and get carried out by the ascending gases and on cooling are deposited in the flues as a solid.

Apart from the potential metal loss there is also the effect of pouring forth vast volumes of sulphurous smoke and dust upon the population and on the surrounding vegetation. 'Smoke farming' in the old days was found by farmers to be a profitable sideline, and smelters had their hands full with the spate of lawsuits which descended on them. In fact it became such a racket that smelters found it cheaper to buy up the surrounding farms and either farm themselves or resell them with a covering smoke clause. The first attempts to deal with the nuisance and recover the fume and dust were by means of brick chambers and zigzag flues, these flues extending in some cases up to many thousands of feet. Thus at the Freiberg Lead Smelter in Germany flues up to 5 miles in length were constructed and in use until the turn of the century, and in Portugal the author has seen flues for the collection of arsenic of extraordinary length zigzagging across the countryside.

In order to provide a larger surface for the deposition of the particles, obstructions in the form of plates and wires were frequently placed in the path of the gases. Freudenberg plates are thin sheet iron plates suspended in the flues at about 3½ in. apart. Fume settles on them and when it has attained a certain thickness falls to the floor of the flue. Rösing wires are suspended vertically in parallel rows from the roof of the flue. Such obstructions were rapidly corroded away. Flues and dust chambers achieved some success, but generally the finer particles refused to settle and escaped into the atmosphere through the chimney stack. To meet this difficulty woollen or cloth bags were introduced.

F. L. Bartlett and G. Lewis, two Americans, were the first to apply cloth filtration on a practical scale. The first bags were 20–30 ft. long and 15–20 in. in diameter and made of wool. These were suspended from the roof of a building, the gas containing fume and dust being introduced through hoppers at the bottom, the gas travelling upwards, depositing the solids on the inside of the bag, clean gas filtering through to an outlet manifold, and being drawn by a fan to the stack.

Automatic bag filters are now used, consisting of cloth filters 8 in. or so in diameter and 12 ft. long enclosed in steel cabinets. Gas containing the suspended solids enters through hoppers at the bottom, passes upwards and through the interstices of the cloth, and deposits its solids, the clean gas being drawn through a manifold at the top to an acid plant or stack as the case may be. The bags are shaken automatically every few minutes, dust being shaken into the hoppers, from where it is removed by screw conveyors. While the bags are being shaken, a current of gas or air is introduced through them in the reverse direction to that normally taken, thus assisting the shaking and cleaning. Alternatively, the bags are emptied by a reverse current of air, the gas current being cut off from one section at a time and a fan used to collapse the bags, the alternate inflation and deflation loosening the deposit.

These automatic bag houses handle up to twenty times the volume of gases formerly handled by the old hand-shaken bags: in other words, one-twentieth of the filtering area is required in the modern type to handle the same volume as the old type.

The automatic bag house is an excellent instrument for the collection of neutral gases, and many thousands of square feet of bag surface are in use; however, when acid gases are encountered, unless the gases are specially conditioned, bag life is extremely short and upkeep is heavy, and for this and other reasons the electrostatic method of dust precipitation is in common use.

ELECTROSTATIC DUST PRECIPITATION

A German named Hohlfeld, in 1824, first discovered accidentally that smoke inside a jar could be dissipated by the introduction of an electrically charged wire. Sir Oliver Lodge some sixty years later rediscovered the phenomenon and constructed a plant on a commercial scale at a lead smelter in North Wales, but the process was not a success, due largely to (a) the limited source of power, for all Lodge had at his command at that time was the high-voltage direct current derived from a Wimshurst machine; (b) no collecting electrode being provided, only a discharge electrode consisting solely of a wire connected to a

Wimshurst electrical friction machine; (c) non-conditioning of the gas.

Dr. F. G. Cottrell of the U.S. Bureau of Mines made an advance in the early 1900s, when the development of alternating current and rectifiers capable of producing 40,000 V or so put the tools in his hand for the production of a successful method for dealing with the smoke nuisance.

The principle of electrical precipitation depends on the fact that a gas becomes charged when passed through a high-voltage field; the electrically charged gas particle tends to move either to the negative or positive electrode, according to its charge. During its movement it collides with other gas particles and also with suspended dust, and moves towards one or the other of the poles dragging along in its wake the charged dust particles. On reaching the collecting electrode the dust gives up its charge and is precipitated thereon. For the discharge electrode, material having a small radius of curvature—such as wire—is used, as it is possible to maintain a high-density charge thereon. For the receiving electrode, materials having a smooth and larger area—such as pipe or plate—are employed. The discharging electrode is connected to a source of high-tension current (30,000 to 60,000 V), the receiving pole being grounded. Direct current has been found to give better results than high-potential alternating current, and the best results are obtained when the discharging electrode is negatively charged, as it has been found that with negative discharge from one-fourth to one-third higher voltage can be maintained between the electrodes than is possible with the positive electrode as the discharging pole.

The apparatus in which the operation is carried out consists of either (a) the pipe type, in which a number of pipes each enclose a centrally spaced wire insulated from the pipes, through which the gases to be cleaned pass either upward or downward or (b) the plate type, consisting of a number of parallel plates with discharge wires evenly spaced between them. The wires in both cases are tensioned by weights. The deposited dust which forms on the pipes or plates is dislodged by an automatic rapping device, the dust falling into bins at the bottom of the precipitator. The plates in the latter type of separator are, when under the influence of high temperatures, apt to buckle and warp and in recent years have been replaced by the 'rod-curtain' type. This consists of a series of iron pipes or rods bolted together to form a curtain or framework so that they can expand freely without buckling and distortion. Wet type precipitators are also in use.

As regards capacity, a smelter turning out say 100 tons of lead per day, would have to handle some 85,000 cu. ft. of gas per minute (from both sintering and furnace operations). To attain an efficiency of 95% a three-unit Cottrell would be necessary, each unit containing 60 to 80 curtains. Operating voltage would be 60–80,000 V.

REFRACTORIES

In those parts of the furnace exposed to high temperature and to the corroding action of the slag, materials capable of withstanding these destructive agencies and at the same time confining the heat must be employed. Such materials constitute refractories. Their essential function is to serve as structural material and thus their usefulness depends on their ability to maintain their mechanical properties at high temperature. Depending on their chemical character they may conveniently be classified as acid, basic and neutral. This division serves as a guide to their suitability for particular applications, since it is generally desirable that refractories should be chemically similar to the slags and fluxes to which they will be exposed in service. The principal members of the three groups are as follows:

Acid: Fireclay, silica and high alumina (bricks of fireclay are generally referred to as firebricks).
Basic: Magnesite, dolomite.
Neutral: Carbon, chrome.

Refractories are mainly produced from naturally occurring rocks and minerals, those produced in the largest tonnages being made from fireclay, for this material is used in some part of practically all furnaces. The chief constituent of fireclay is the silicate mineral kaolinite ($Al_2O_3 \, 2SiO_2 \, 2H_2O$) the composition of which is SiO_2 46·3%, Al_2O_3 39·7%, H_2O 13·9%. Generally, however, the SiO_2 ranges from 50–70% and the Al_2O_3 from 20–40%. Fireclays may hence be regarded as mixtures of pure china clay (kaolinite) with excess of silica and small quantities of earthy silicates. The fired product contains a proportion of mullite, a highly refractory form of aluminium silicate.

In Great Britain fireclays occur widely among the rocks of the coal measures and firebrick manufacture is accordingly carried out in many coal-producing areas.

The high aluminous refractories can be used under more severe conditions than ordinary firebricks. They contain 50% Al_2O_3 and over, and are made from a number of different materials such as kyanite ($Al_2O_3 \ SiO_2$), sillimanite ($Al_2O_3 \ SiO_2$) and andalusite. These minerals on firing yield high proportions of mullite.

Silica ranks next in volume of production and materials of this class consist largely of a fine grained quartzite rock known as ganister containing 98–99% SiO_2. It occurs in the lower coal measures of south Yorkshire, and is largely worked in the neighbourhood of Sheffield, which is one of the main centres for the production of silica refractories.

Basic refractories are composed mainly of magnesite ($Mg \ CO_3$) and dolomite, the former being the most widely used. For metallurgical use it is necessary to heat the magnesite before it can be used owing to the large shrinkage which the mineral undergoes on firing, and hence it is essential to reduce it to a condition in which the volume will remain constant. For this reason it is 'dead burned' in a kiln at about 1700° C., the resulting MgO sintering together to form a dense hard crystalline structure known as Periclase.

Dolomite, a double carbonate of magnesium and calcium, is an important lining material for basic Bessemer converters and for open-hearth furnace bottoms.

Chromite refractories are employed in furnaces as partings between the silica and magnesite bricks and also in combination with magnesite as roof material in the basic open-hearth furnace. For this purpose they are replacing the conventional silica brick, at least one-third of the open-hearth furnaces in this country being now so roofed.

Carbon. Refractories of this type are manufactured by firing a mixture of powdered coke or graphite with tar or pitch. They are used in the hearths of blast furnaces and also in the manufacture of crucibles employed in the melting of metals.

71

HISTORICAL DEVELOPMENT

The early history of refractories is related to the history of pottery, but the history of refractories during the past century or so is concerned with the growth of the metal industries, since these could advance only as satisfactory refractories became available for the construction of furnaces.

Records indicate that mica schist was in use as a lining for the early (iron) blast furnaces. This material which is largely composed of silica was roughly dressed and shaped into blocks before being placed in position. Fireclay became available for furnace linings in the 18th century.

The advent of man-made silica brick refractories occurred later than the fireclay materials and was the result of a demand for a refractory of better heat-resisting quality than firebrick. They were first manufactured in England by W. W. Young, a maker of porcelain who made them from the Dinas rock and sand found in the Vale of Neath in South Wales. Dinas rock consists essentially of silica (97% SiO_2) with small quantities of alumina, iron and lime. Young manufactured silica bricks by coarsely crushing the rock, mixing with a little lime or clay and water and then pressing into iron moulds by a lever press worked by hand; drying and heating in a small kiln for about a week completed the process. Young was also instrumental about 1858 in commencing the manufacture of ganister brick in Sheffield on the site of the present Oughtibridge Silica Firebrick Co. Ltd. Present-day manufacture follows the same general lines except that the process is now highly mechanized. The material is ground and screened to reduce it to a suitable size, then mixed and tempered with water and lime, the mixture being moulded by power press, dried and heated in oil-fired kilns. Continuity of operation has been achieved by loading the bricks on cars and passing through long tunnel kilns. A recent development in power press technique is the removal of entrapped air from the ground clay by pressing the bricks under vacuum. This in addition to increasing the density, reduces porosity and permeability, thereby improving spalling and resistance to slag attack.

Basic Refractories. Use of basic refractories dates back to the invention of the basic-lined Bessemer converter in 1877 by

Gilchrist and Thomas who used rammed dolomite linings to remove phosphorus from molten steel. On the introduction of the basic open-hearth process, dolomite and magnesite hearths were used for the same purpose.

The first large-scale use of magnesite in the lining of open-hearth steel furnaces in the U.S.A. dates from 1886 when the Otis Steel Company, Cleveland, changed the bottom of its open-hearth furnace from siliceous to a basic bottom consisting of magnesite imported from Radenthein in Austria. During the First World War when supplies were cut off, magnesite was mined in California and Washington and later magnesia was produced from oceanic salts by the Dow Chemical Company, Midland, Michigan. No natural deposits of magnesite occur in Great Britain and before 1938 all 'dead burned' magnesite was imported from Austria, Greece and India. During that year the successful development of a process for the extraction of magnesia (MgO) from sea water led to the erection of a large-scale plant by the Steetley Co. Ltd., at West Hartlepool, Co. Durham. This plant has a capacity in excess of 150,000 tons per year.

Chrome-ore refractories were first used on a large scale in the 1880s in the furnace of the Petersburg-Alexandrofsky Steelworks in Russia, and in 1896 were being used in the U.S.A. as a neutral parting to separate magnesite and silica, these refractories tending to react chemically at high temperatures. In 1918 chrome brick was used in copper-refining furnaces, but their use has somewhat decreased during the last few years due principally to spalling tendencies.

The use of oxygen in open hearth furnaces has caused the adoption of chrome-magnesite bricks in place of the conventional silica roof. Roof life with silica brick using oxygen was extremely short, replacement with chrome-magnesite brick resulting in consistently longer roof life. The bricks consist of crushed and ground chromite ($FeOC_2O_3$) and dead burned magnesite bonded together with a little lime. The mixture is placed in a mould and consolidated under heavy pressure. The bricks are then dried and fired in a kiln.

Alumina itself with its high fusion point is an excellent refractory and the range of high aluminous materials have in recent years been considerably extended. A feature of this development has been the increased use of calcined bauxite

73

from British Guiana and the exploitation in the U.S.A. of the Missouri diaspore ($Al_2O \cdot 3H_2O$) deposits. This mineral is mined with its associated clay matrix and shows an alumina content after calcination of approximately 70%.

BIBLIOGRAPHY

1. 'Seventy-five Years of Progress in the Mineral Industry (1871–1946)', A.I.M.E., New York.
2. LIDDELL, D. M. (Editor), *Handbook of Non-Ferrous Metallurgy*, 1926. McGraw-Hill Book Co., New York.
3. GOWLAND, W., *Metallurgy of the Non-Ferrous Metals*, 1918. C. Griffin & Co. Ltd., London.

CHAPTER FOUR

IRON AND STEEL

ONE hundred and twenty years ago Britain, then the largest producer in the world, made approximately 60,000 tons of steel and 3 million tons of pig iron, about a half being converted to puddled (wrought) iron. Iron was thus the principal metal on which the Industrial Revolution had been based, steel being a comparatively rare metal made by arduous and expensive methods.

In 1960 British production of steel totalled 24 million tons, whereas only 50,000 tons of puddled iron was manufactured.

The modern steel age as we know it was born with Bessemer's discovery in 1856 of the rapid conversion of pig iron into steel without the use of fuel. The eventual success of this process coupled with the increasing demand for steel for railway tracks —by far the largest single consumer—attracted other inventors; of these the open-hearth process invented by C. W. Siemens in 1862 was destined to become a rival to the Bessemer process, and in fact in 1894 outstripped the total tonnage produced by the Bessemer furnace, the relative figures being open hearth 1,575,000 tons, Bessemer 1,535,000 tons. Production of puddled iron in that year was 1,339,000, amounting to not much more than 15% of total pig iron production. Since then the output of open-hearth steel, both here and in the U.S.A., has far outstripped that of Bessemer steel, the main reasons being the ability of the former to use large quantities of steel or iron scrap and that it places no restrictions on the type of raw material used, iron of any phosphorus content being capable of utilization.

IRON

The primary product produced by smelting iron ore in a blast furnace is pig iron. A majority of the pig iron nowadays is conveyed in a molten condition to open-hearth furnaces or

75

converters and refined to steel. About 15% (in Great Britain) of the pig iron is allowed to solidify, and is then remelted and run into moulds, the product being cast iron.

Wrought Iron. In the mid-1800s, of the pig iron production which amounted in Great Britain to about 3 million tons, approximately one-half was cast into moulds designed to the shape required in the final article. The other half was converted into a form known as wrought or puddled iron. Wrought iron was the first form of iron known to man and for centuries remained the only variety available. It is a malleable type of iron which could be hammered into the various shapes required by primitive man for his weapons and tools. It was produced from oxide iron ore and charcoal in shallow hearths or shaft furnaces, a hole being provided at the base of the furnace through which a blast of air was induced naturally by the wind and later by bellows. The product was a spongy mass of iron intermixed with slag. This was reheated and hammered to expel the slag and then wrought or forged into the desired shape. The mass of wrought iron was referred to as a 'bloom', the furnaces being known as bloomeries.

Attempts to speed up the process by an increase in the working temperature usually resulted in failure, for the product instead of being malleable proved to be hard and brittle, the metal fracturing on attempting to forge it. The product was in fact pig iron, the high temperature causing carbon to be absorbed from the charcoal, and silicon from the gangue; these impurities adversely affected the properties desired by the ancients.

In order to utilize the high-carbon product for making forged or wrought articles it was necessary to develop processes that would remove the excess carbon and silicon from the iron to produce the desired soft malleable wrought product. Eventually a method was evolved known as the puddling process, and this became universally adopted.

Puddling Process. The early method for the production of wrought iron from pig iron consisted in remelting the pig in a hearth in contact with charcoal fuel, the metal being decarbonized and desiliconized by the oxygen in the air blast. The pasty mass of iron was then remelted, the subsequent hammering of the bloom expelling much of the slag and charcoal. This was in essence the Wallon process introduced into Sweden* by the Belgians about 1600. An alternative process was also introduced into Sweden known as the South Wales process because it was operated by Welshmen. This was a two-stage method, usually conducted in a two-hearth furnace, the upper being used for melting the pig, the final refining taking place in the lower.

In 1783 Henry Cort introduced his puddling furnace, its essential advantage being that coal and coke could be used as fuel and no blast was required. Previously the sulphur in these fuels had prevented their use, the sulphur imparting embrittlement to the metal. Further, the rapid depletion of forests to supply wood for charcoal was halted.

The furnace was of the reverberatory type, the essential feature being that the hearth had been hollowed out so as to make a puddle of molten iron. A firebox about 3 ft. square was separated from the hearth or puddling basin by a bridge. The hearth, measuring 6 ft. in length by 4 ft. wide, was of firebrick lined with sand, which in operation, was rapidly fluxed away by the iron oxide formed by the partial oxidation of the pig iron to form iron-silicate slag. The process was wasteful of iron, the losses in the slag amounting to about 30% of the metal charged. The substitution of old hearth material consisting of iron silicate containing iron oxide in place of the sand bottom, initiated by J. Hall in 1820, not only increased the yield to about 90%, but shortened the time of the heats by accelerating the decarburizing process. The new process was called 'wet puddling' because of the formation of a large amount of slag; the older 'dry puddling' method produced much less slag.

The raw material consisting of 4–5 cwt. of pig iron was charged by hand to the reverberatory and heated to 1100° C., constant rabbling being necessary in order to expose it fully to the flames and to prevent the material sticking to the hearth. Oxidation of the silicon and manganese took place, the oxidation being induced by addition of iron oxide scale. Carbon was the last to be oxidised, carbon monoxide being formed ($FeO + C = Fe + CO$) which escaped as bubbles bursting into flame known as 'puddlers candles', this stage of the process being

---

* Sweden (which has long been known for the high quality of its wrought iron) was chosen because its ore was low in silicon, sulphur and phosphorus, and owing to the constant replanting of the forests ample supplies of charcoal were assured.

known as the boiling stage. When evolution of CO has ceased, the metal due to its increase in melting point becomes pasty and very hard to work; in the phraseology of the puddler, the iron 'comes to nature'.

The temperature was then raised and the puddler rabbled the metal to eliminate as much slag as possible, and then worked the pasty iron into a number of balls weighing roughly 80–100 lb. each. The balls were then removed by tongs and transferred to a steam hammer (invented in 1839) for removal of surplus slag, the bloom then being rolled into bars. The operation required about 2 hours, 5–6 heats being achieved in the twelve-hour shift at a consumption of 20–25 cwt. of coal per ton of iron. The work was very arduous and the pay poor.

Percy* described the puddler and his 'underhand' (helper) naked to the waist, taking turns constantly to stir the bubbling 4 cwt. charge with their iron rabbles until it 'came to nature' and the 80 lb. balls were lifted out of the furnace. It was he said 'probably the most severest kind of labour in the world'.

Puddling developed into an enormous industry in the Midlands, especially in the Black Country where underlying coal seams 10 yards thick and ironstone with a metallic content of 35% and more stretched from Birmingham in the east to Wolverhampton and Stourbridge in the west. Wrought iron production from this district rose to 670,000 tons in 1873, the number of puddling furnaces being 2,100. South Wales was another big centre for puddling furnaces, the wrought iron being rolled into sheets for tinplate manufacture, in which Wales, by 1860, had established a virtual monopoly.

Mechanical Puddling. From about 1860 onwards almost incessant research went on to devise a mechanical substitute for the incredibly hard manual labour involved in puddling. At first these mechanical devices took the form of stirring or rabbling appliances that could be attached to the top of the furnace. The more successful attempts involved a complete change in the design of the furnaces. Menelaus in 1865 experimented with a rotating furnace at Dowlais and S. Danks of Cincinnati, U.S.A., built a cylindrical furnace that rotated about a horizontal axis, and by 1872 there were 74 Danks furnaces being built in England. Wrought iron, however, had

* J. Percy, *Metallurgy of Iron and Steel*, 1864, p. 656.

begun to pass its zenith, 1875 heralding the decline in production of puddled iron; steel production was then beginning to gather momentum, total production of puddled bar exceeding 3 million tons in 1873, whereas in 1877 it had fallen to 1·8 million tons. Bessemer-steel production rose from 215,000 tons in 1870 to 835,000 tons in 1879 and open-hearth steel from 88,000 tons in 1875 to 137,000 tons in 1877 and 175,500 tons the next year.

Aston-Byers Process. In the U.S.A. a radical departure in manufacturing puddled iron had been achieved. In the Aston process, metal refining, slag melting and balling are carried out in separate stages, each stage being operated in a separate furnace. Pig iron is blown to almost pure metal in an acid-lined Bessemer converter and then poured at a controlled rate into a bath of iron silicate molten slag held below the melting point of iron. Reaction takes place, at the end of which the iron 'comes to nature', the huge balls weighing 5–10 tons being electrically pressed for removal of slag, and then rolled into slabs or billets. The slag, which is separately prepared by heating silicon materials and iron ore, acts as a heat-absorbing agent effecting solidification of the metal.

IRON

The raw material used in the production of wrought iron described in the previous section is pig iron which is also the raw material for the production of steel. The operation involved in the manufacture of pig iron from its raw materials—ore, coke, limestone and air—is performed in a blast furnace which produces molten high-carbon iron and a slag.

The majority of improvements which have resulted in the modern blast furnace—towering 100 ft. or so into the sky and with a daily output of up to 2,000 tons of pig iron—had almost all been made by the mid-19th century.

One of the earliest and most important developments had taken place in Shropshire when Abraham Darby (1677–1717) in 1709 substituted coke for charcoal. This simple step opened up the iron industry to the industrial revolution, for coke being hard and porous is able to withstand a far greater burden without crushing, thereby making possible the construction of much larger furnaces with a resultant increase in output. It also eased the situation caused by the depletion of the forests. The crude coke produced at that time, however, had certain disadvantages. It possessed a higher ash content than charcoal, more phosphorus and generally more sulphur, and consequently the iron was less pure and more brittle. While not suitable for refining into wrought iron, the pig was, however, suitable for castings.

It was found that the use of coke necessitated a more powerful blast than that obtainable from water-driven bellows. About that time (1774) John Wilkinson, a Shropshire ironmaster, had produced a satisfactory method for the more accurate casting and boring of cast iron cylinders, and with the aid of James Watt the first steam-driven blowing engine was used to blow Wilkinson's furnace in Shropshire. In Britain by 1806 there were 162 coke blast furnaces and only 11 charcoal furnaces. The country's pig iron production in that year was some 258,000 tons.

The next important step was the preheating of the blast. Until 1828 air blown into the furnace was at its natural temperature. It occurred to a Scots engineer, J. B. Neilson (1792–1865), that a reduction in the consumption of fuel in the furnace might be obtained by preheating the blast. He obtained a patent in 1828, his ideas being first applied at the Clyde Iron Works. The first hot-blast apparatus of Neilson was very simple, consisting of an iron stove 4 ft. × 2 ft. × 3 ft. set in brickwork and heated by a fire below. The cold blast entered the stove above the grate and passed out from the opposite end directly into a blast furnace tuyère, having attained a temperature of about 95° C. Each tuyère of the furnace was equipped with one stove. The effect of even this moderate preheat was remarkable, coal consumption per ton of iron dropping from 8 tons to 4–5 tons, the production rate of metal increasing nearly 100%. It was soon found, however, that due to corrosion caused by the heat, the stoves had a very short life. To combat this, Neilson replaced the stove by an arrangement of cast iron pipes fitted horizontally over the grate and enclosed in a refractory chamber. This arrangement raised the temperature of the blast to such a degree (300° C.) that it caused rapid deterioration of the iron tuyères and led to the invention of the water cooling of the tuyères by J. Condie at the Blair ironworks in Scotland.

The fuel economy made possible by the hot blast attracted much attention and led to the adoption of a device whereby the hot furnace gases which issued from the top of the blast furnace and previously had simply been burnt, were applied to the task of heating the incoming cold blast. In the year 1809 blast-furnace gases began to be employed in France for the purpose of brickmaking and for heating small furnaces. Attempts to utilize this gas as a source of heat regeneration were made in Germany in the 1830s, but the first successful utilization of waste furnace gas to heat the blast was in South Wales in 1845, the gases being taken off the top of the furnace and conveyed through pipes to the hot-blast stoves, where they were burnt to heat the cold blast. The year 1857 saw the introduction of the Cowper hot-blast stove which is still in universal use today. E. A. Cowper—a friend of Frederick Siemens—applied Siemens heat regenerative principle in a system of refractory bricks set checker-fashion in a tall cylindrical steel casing, an internal vertical partition dividing it into a combustion chamber provided with a burner and a second section composed of refractory brick checkerwork consisting of a multiplicity of bricks arranged to expose a maximum of surface. Blast-furnace gas together with air is admitted via the burner, combustion taking place, the products of combustion passing upwards and over the partition and then downwards through the checker-work to which they yield up their heat. The gas supply is then periodically shut off and cold blast air admitted at the bottom and allowed to pass through the heated checkerwork before being admitted to the furnace. Thus the cold blast and heated gas alternate through the stoves, the burned gas giving up its heat to the bricks which in turn give up heat to the blast.

Cowper's stove raised the temperature of the blast air to 600° C. thereby substantially increasing the output of pig iron. He originally used blast-furnace gas direct from the top of the furnace to preheat the stove; but the dust with which the gas was laden choked the stove checkerwork, and extensive dust-catching equipment had to be used before satisfactory results were obtained.

The blast furnace was at that time closed by a device invented in 1845 by J. P. Budd. The arrangement known as the 'cup and cone' consisted of a cone-shaped hopper, the smaller opening of which was closed by a cup or bell that could be raised or lowered at will. With the cup raised against the hopper, the raw materials were dumped into the hopper, the lowering of the bell causing the charge to fall rapidly into the surface. As large quantities of gas escaped with each lowering of the bell, this device was later improved by the addition of a second but smaller bell and hopper above the first, thus providing a gas-tight space between the two. The raw material is first dumped into the upper hopper, where it falls into the large hopper below, when the small bell is lowered. The small bell is then raised to seal the upper hopper, the large bell is lowered, the charge falling into the furnace without any escape of gas. This device also ensures a more even distribution of the charge.

EVOLUTION OF THE BLAST FURNACE

The old charcoal furnaces of the 17th century, usually between 12 and 18 ft. high, were stone-built structures, rectangular in cross-section, often built against a hillside from which raw materials could be charged into them by wheelbarrows via a gangway. Apart from an increase in height there was little change in blast-furnace construction until the advent of the hot blast, which to increase production led to a greater number of tuyères and to important alterations in the cross-section of the furnace. The furnace was given a cylindrical shape, and made of cast iron plate lined on the inside with refractory firebrick, the whole being supported by cast iron columns allowing of easy access to the tuyère region. Furnaces of this type were known as Scotch furnaces because they were first erected in Scotland in the early 1830s (Fig. 5). They became the standard type and laid the foundation for the modern type of blast furnace. In Britain between 1850 and 1870 the stack height was progressively raised from 40 ft. or so to 80 ft., the diameter, however, being kept on the small side owing to the reluctance of blast-furnace designers to develop a non-penetrative area at the tuyère zone. Outputs for the 80 ft. furnace were of the order of 350 tons per week.

It was to the U.S. that the main initiative now passed. By increasing both the hearth diameter and the blast pressure substantial increase in tonnage took place. One of the first of these furnaces with an increased hearth area was the Isabella furnace located near Pittsburgh. Dimensions were 75 ft. high, diameter 20 ft. and capacity 15,000 cu. ft. The furnace was in operation from January 1876 until May 1880 and made a total of 119,000 tons or an equivalent of 2,300 tons per month with a consumption of 3,153 lb. of coke (per ton of iron) and a blast temperature of 600° F. A similar furnace at the Pittsburgh works of Carnegie Phillips & Co. produced on an average 3,338 tons per month in 1878 with a coke consumption of 2,749 lb. per ton. With the increase in the size of furnaces, the transportation of the ore, flux and fuel by means of wheelbarrows became

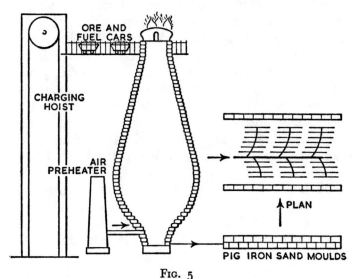

FIG. 5

IRON BLAST FURNACE c. 1840

impossible, these hand methods being superseded by mechanical traction and handling; the raw materials were raised to the top in skips running on inclined tracks extending from the top of the furnace to the raw material bins at the foot of the furnace.

The invention of equipment to determine the level of the charge within the furnace by D. Baker and its installation at the Illinois Steel Co. in 1901 gave the operator his first real knowledge of the flow of material through the furnace.

Before 1900 iron was tapped from the furnace and conducted along channels in the ground to a series of sand moulds known as pig beds owing to a fancied resemblance to a nursing litter of pigs, the main channel being known as the sow and the branches as pigs. When the iron became cool, the pigs were broken from the sows into pieces weighing about 1 cwt. With increase in the output of blast furnaces, resort was made to a casting machine consisting of a number of moulds carried on an endless chain passing over a head and tail sprocket wheel. The molten iron is poured into the moulds from a ladle and cooled by water sprays as the chain travels to the head sprocket, falling from the mould into cars or wagons. The method was patented in 1894 by Uehling.

By the close of the century all the major improvements had been made that determined the general line of process for the manufacture of iron in the blast furnace.

One of the interesting facts in a study of blast furnaces during the last hundred years is that the increase in tonnage is out of all proportion to the increase in size of the furnace, technical progress, construction and operation having increased the total blast-furnace tonnage with far fewer furnaces. A century ago output per blast furnace in Britain was some 9,000 tons of pig iron per annum, at an expenditure of 3 tons of fuel per ton of iron, whereas today 200,000 tons is produced at a fuel rate of 0·8:1.

Although simple in construction the present-day iron blast furnace (Fig. 6) is a highly efficient apparatus which operates continuously producing 1,500–2,000 tons of pig iron per 24 hours with a metal recovery of 98%. It is in effect a vertical circular shaft between 80–100 ft. high, the shaft increasing in cross-sectional area from the top or throat downwards, a maximum being attained at the bosh some 25–30 ft. from the ground. At the throat it may be between 15–20 ft. in diameter, at the bosh between 20–30 ft. and at the hearth 15–20 ft. The furnace, which may weigh up to 12,000 tons, is provided with a massive concrete base in which are embedded steel columns upon which rests a horizontal heavy steel rim. Known as the mantle, it supports the furnace stack in such a manner that the hearth and bosh brickwork may be removed or renewed without disturbing the stack proper.

The hearth is situated at the bottom of the furnace for receiving the molten iron and slag. The base of the hearth is composed of 7–10 courses of firebrick giving a total thickness of 9–12 ft. over which are lain two layers of carbon blocks. The sides of the hearth are constructed of two courses of carbon brick of a total thickness of 3 ft. Located about 3–4 ft. above the surface of the bottom of the hearth is the iron notch through which the molten iron is periodically tapped off. The slag which floats on top of the iron is withdrawn through the cinder notch situated above the iron notch but below the tuyères. When not in use the tapholes are plugged with clay. In time the bottom is eroded being replaced by a layer of metal known as a salamander.

Immediately above the hearth the walls of the furnace extend

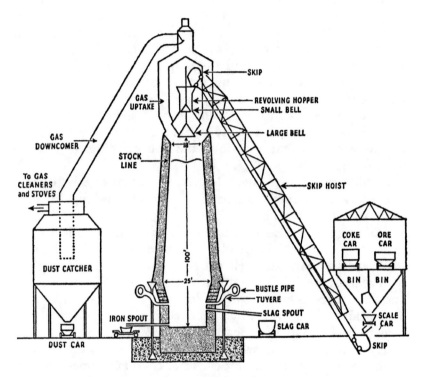

FIG. 6

SECTIONAL VIEW OF MODERN IRON BLAST FURNACE AND ANCILLARY EQUIPMENT

outward to a maximum diameter at the mantle, the bricks being stepped out for a height of 9–12 ft. and at an angle of 80–85° to the horizontal. The bosh serves to retard and to compensate for the volume reduction in the charge as it melts and descends into the smelting zone in front of the tuyère, the hottest part of the furnace. The retarding of the charge serves to intensify combustion and hence leads to a more rapid fusion. In order to cool and protect the refractory bricks, hollow metal plates through which water is circulated are inserted into the brick-work.

The upper part of the furnace arising from the bosh comprises a tall vertical shell lined with refractory brick varying in thickness from 2–5 ft. The refractories are usually a dense fired aluminous brick 35–45% alumina content possessing a life of up to six years.

Immediately above the hearth are situated the tuyères, circular orifices through which the air required for combustion is blown. According to the size of the furnace they number 10–20, being spaced around the diameter of the furnace and are composed of copper, and water cooled. Air reaches the tuyères via a bent pipe known as a gooseneck from a large refractory-lined main or bustle pipe which encircles the furnace. A peephole is usually located in the tuyère to permit the inspection of that portion of the furnace charge directly in front of the tuyère.

Ancillary equipment includes plant external to the furnace, essential to such operations as raw material handling, air pre-heating and gas cleaning.

The raw materials—iron ore, coke and limestone flux—are stored in bins near the base of the furnace, being fed from large stockyards by cars and belt conveyors. A scale car equipped with scales to weigh its contents draws material from the bins and discharges its load to skips holding about 5 tons of material. These skips are drawn up an inclined ramp to the top of the furnace and dump their load in the receiving hopper of the double bell feeder from whence the material enters the furnace upon lowering of the large bell. Ore, coke and limestone are usually charged in rotation, the relative weights varying according to the composition of the ore.

The air required for combustion is preheated before its admittance to the furnace through the tuyères. This serves to intensify and speed up the burning of the coke with a resultant economy in the use of fuel. Preheating is carried out in Cowper stoves (page 81). Blast-furnace gas is admitted via a burner to the combustion chamber where it ignites, the products of combustion pass upward and then over into the checker chamber, heating up the bricks and then out at the bottom of the chamber to the stack. The cold blast, compressed by the blowers, passes in at the bottom of the checker chamber, travels upwards, absorbing heat and then over the top and downwards through the combustion chamber and out into the hot-blast bustle pipe, having been preheated to a temperature of 600–800° C. Usually three (or four) stoves are provided, one stove being 'on blast', the others being 'on gas', the change-over from gas to blast and vice versa being automatically performed by valves.

Modern stoves are up to 30 ft. in diameter, and 70–100 ft. high and contain 500 tons of bricks which provide 200,000–250,000 sq. ft. of heating surface.

The gas issuing from the top of the blast furnace is heavily laden with dust varying in size from $\frac{1}{4}$ in. down to a few microns, and arrangements have to be made to remove this dust to prevent choking of the stove checkerwork. This is usually accomplished in two or three stages, the primary stage consisting of precipitation of the dust by reducing its velocity—as it is carried over in the downcomer from the furnace top—in a large brick-lined chamber known as a dust catcher, which allows the coarser particles to settle out. Further cleaning is effected by water washing, cyclones and electrostatic precipitation.

In addition to heating the stoves, the heat value of the gas is utilized in the air blowers, for generating steam and electricity and for firing coke ovens and furnaces in adjoining steel works.

Blast-Furnace Reduction Process. The main reagent in the reduction of iron oxide to metallic iron is carbon monoxide generated by the oxygen in the air blast coming into contact with the hot coke in the tuyère zone. Carbon dioxide is first formed

$$C + O_2 = CO_2 + 14{,}550 \text{ B.T.U./lb. C.}$$

which in the presence of excess carbon is reduced to carbon monoxide.

$$CO_2 + C = 2CO - 5{,}730 \text{ B.T.U./lb. C}$$

the result being

$$2C + O_2 = 2CO + 8{,}820 \text{ B.T.U./lb. C.}$$

In considering the reduction in the descending ore charge,

87

it is convenient to distinguish four zones, the zone of preparatory heating, the upper zone of reduction, the lower zone of reduction and fusion and the zone of separation.

1. Zone of preparatory heating. Inside the furnace as the charge sinks down and comes into contact with the ascending hot gases at a temperature of about $150°$ C., moisture is lost and the mass becomes porous.

2. Upper zone of reduction. Reduction of the iron oxides commences at a temperature of $400-500°$ C., some 65 ft. above the hearth, the reactions which occur being as follows:

$$3Fe_2O_3 + CO = 2Fe_3O_4 + CO_2 + 48 \text{ B.T.U./lb. } Fe_2O_3$$

$$Fe_3O_4 + CO = 3FeO + CO_2 - 68 \text{ B.T.U./lb. } Fe_3O_4$$

$$FeO + CO = Fe + CO_2 + 95 \text{ B.T.U./lb. } FeO$$

3. Lower Zone of Reduction. At about $800°$ C. decomposition of limestone occurs

$$CaCO_3 = CaO + CO_2$$

the lime combining with some of the gangue material to form slag. Manganese, silicon and phosphorus oxides are reduced by carbon;

$$MnO + C = Mn + CO$$

$$SiO_2 + 2C = Si + 2CO$$

$$Pb_2O_5 + 5C = 2P + 5CO$$

Of the manganese about $50-75\%$ alloys with the iron, the remainder passing out with the slag. Although the greater part of the silica combines with the lime and other bases to form silicates, that which is reduced to silicon combines with the iron to form iron silicide. All the phosphorus is absorbed by the iron, very little being eliminated in the slag. Sulphur is largely removed as the iron passes down through the molten slag,

$$FeS + CaO + C = CaS + Fe + CO$$

some sulphur, however, always remaining dissolved in the molten iron. Carbon to the extent of $3-4\%$ is also absorbed by the molten iron.

Carbon alone also acts as a reducing agent.

$$FeO + C = Fe + CO$$

All reductions are complete at about 1250° C.

4. The slag and the iron both now in the liquid state seep down through the coke to the hearth, the iron and slag separating into two layers. The iron being the heavier, sinks to the bottom and is tapped every six hours or so, flowing into a refractory-lined ladle and being then transported to the steelworks. Slag is tapped from the slag notch and runs into ladle cars for transport to the dump.

Products. A balance sheet of the materials concerned in the production of one ton of pig iron is as follows:

Furnace Charge	Tons	Products	Tons
Iron Ore	2·0	Pig iron	1·0
Coke	0·8	Slag	0·7
Limestone	0·4	Gas	5·4
Air	4·0	Dust	0·1
	7·2		7·2

Pig Iron. The amounts of phosphorus and silicon present in the ore largely determines the grade of iron. Basic iron has a high phosphorus and low silicon content, but hematite iron is low in phosphorus and high in silicon. Most of the iron produced in Great Britain is of the basic type. Typical assays are:

Basic: Si 0·95%. S 0·04%. P 0·54%. Mn 1·4%. C 4·0%
Hematite: Si 1·82%. S 0·03%. P 0·04%. Mn 0·8%. C 3·7%

Slag. Blast furnace slag consists essentially of calcium and aluminium silicates ranging in composition from 30–35% SiO_2, 35–40% CaO, 5–10% MgO and 12–20% Al_2O_3. The amount of iron lost in the slag is primarily a function of operational efficiency, but in general does not exceed 0·5%. Slags find some use in industry as an aggregate for roadmaking and for cement manufacture.

Gas. The gas leaving the top of the furnace (which amounts to approximately 5 tons per ton of pig iron produced) is composed of 25–30% carbon monoxide, 10% carbon dioxide, 1–3%

hydrogen, the remainder being nitrogen. It possesses a calorific value of 90 B.T.U. per cu. ft. or about 1/5 that of town gas.

Dust. Gas leaving the furnace carries with it 10–20 grains of dust per cubic ft. amounting to approximately 10% of weight of the pig iron produced. It is composed of SiO_2, Fe_2O_3, CaO, C and Al_2O_3; after collection in the dust catchers and gas cleaning plant it is pugged with water, fed to the sintering machine and hence finds its way back to the furnace.

Recent Innovations. Operational improvements in recent years have resulted not from increase in size of the furnace, but mainly as a result of:

(*a*) Better physical preparation of the raw material.
(*b*) Reduced thermal requirements through the use of higher grade ore giving lower slag volumes.
(*c*) Increased heat input through a higher hot-blast temperature.
(*d*) Use of high top pressure.
(*e*) Tuyère injection.

The output and efficiency of a blast furnace is intimately connected with the physical properties of the material charged, for this has an important bearing on the upward flow of the reduction gases through the charge. If the material is too fine, the stack becomes choked and the flow of gases is impeded. On the other hand if the material is too coarse, the gases will not have had sufficient time to carry out reduction before the ore on its downward travel reaches the hearth. Hence, efficient crushing and screening of the charge to effect a uniform-sized product is essential. Modern practice tends towards a complete preparation of the furnace burden and in many cases this includes sintering all the ore; the porous character of the material makes it more readily reducible by the furnace gases with a resultant reduction in coke consumption. The advantages are such that in Great Britain sinter is now providing about 18 million tons (50%) of the iron charged to the blast furnace; there having been a sevenfold increase since the last war. Similar trends are also evident in other countries; in Sweden for example, many of the blast furnaces work with 100% sinter.

High Top Pressure. In normal practice the gas leaves the blast furnace at a pressure of about 2 lb. per sq. in. By installing a throttling valve at the furnace top an increase in the gas pressure to 10–15 lb. is possible. The advantages are:

1. The increase in pressure generated inside the furnace compresses the gas, thus increasing its density and permitting more air to be blown into the furnace without increasing the gas velocity; the output of the furnace thereby increasing.

2. For a given flow of gas up the furnace the lifting power is reduced and the carry-over of dust is decreased. Since dust carry-over can be a limiting factor in the rate of driving, higher pressures permit greater driving rates.

3. The increase in density of the air results in a reduced resistance to its flow and this leads to less channelling and a better gas-solid contact which in turn permits a reduction in the coke consumption.

Operating with a top pressure of 10–15 lb., it has been found possible to increase iron production 10% with a reduction in coke consumption of 5% and a decrease in the amount of flue dust of about 20%. The operation of blast furnaces under elevated top pressure began in America in 1943, and is now adopted as routine practice in many large iron works.

Tuyère Injection. 1. Oxygen. The air blown to the air blast furnace consists of 21% oxygen and 79% nitrogen. By using an oxygen-enriched blast the percentage of the inert constituent nitrogen is reduced which has the effect of intensifying combustion and increasing the amount of carbon monoxide available for reduction. Ironmaking capacity is thus increased and the merits are such that the Steel Company of Wales and others have adopted this technique. In Russia the use of oxygen-enriched blast together with injection of natural gas has resulted in not only an increase of production of more than 30% of pig iron, but also in an appreciable economy in the use of coke.

2. Fuel Oil. It has been established that injection of fuel oil through the tuyères results not only in an increase in ironmaking capacity, but also a saving in fuel. Thus injection of 5 cwt. of oil plus 3,000 cu. ft. of oxygen per ton of metal has resulted in a coke rate of only 6 cwt. per ton of iron with an increase in output

of some 10%. It is to be noted, however, that because of poor utilization of hydrogen in the blast furnace, hydrocarbon fuels can only be used in conjunction with oxygen.

3. Coal. Trials are now taking place at the Stanton and Staveley Ironworks and in Buffalo in America involving the injection of low grade coal in pulverized form. The coal is crushed and injected with the preheated air blast at about 11 p.s.i. and at a rate of about 16 lb. per minute. A reduction of the normal amount of coke of nearly 20% has been achieved with no adverse effect on the quality of the iron produced.

The inducement to save coke has been the cost of British coking coal, which has more than doubled in the last ten years. As an index of the saving achieved it can be pointed out that in 1920 28 cwt. of coke were required to make one ton of iron; the average is now 16 cwt., and a further marked reduction is to be expected.

STEEL

The primary product produced by smelting iron ore in a blast furnace is pig iron. The transformation of pig iron into steel arises mainly as a result of the reduction of the carbon content from 3·5% to below 1·0%.

The approach to making steel in ancient times consisted not in a reduction of carbon but in an increase in the carbon content of the raw material which was wrought iron, by heating in contact with carbon, the latter diffusing into the iron. This method came in time to be known as the cementation process. A further method, known as the crucible process, consisted in melting wrought iron in clay crucibles in presence of carbon. Both of these methods were practised by the ancients.

Cementation Process. The first approach to making steel in more modern times occurred in 1614 when a patent was granted to Meysey and Elliot in England for the manufacture of steel by the cementation process. This cementation process was practised in the Sheffield area and elsewhere and was in use until quite recent times.

It was operated in a furnace (Fig. 7) shaped like a pottery kiln with a conical brick stack some 40 ft. high to radiate heat from the waste gases and effect a crude heat conservation. Because of its extreme purity the raw material was usually Swedish bar iron, 6 ft. long, 3 in. by ¾ in., which was placed in furnace pots built up of sandstone above the firegrate. The bars alternated with layers of lump charcoal each pot holding up to 20 tons. The pots were then covered and made airtight by luting down the lid with clay. Firing at a temperature of 900–1100° C. continued for 8–12 days. When the desired temper or hardness had been reached (ascertained by the fracture of a trial bar withdrawn from a pot) the fire was withdrawn and the furnace left to cool for a week. The bars were then sorted according to the appearance of the fracture, the latter indicating the amount of carbon absorbed. Carbon content ranged from 0·7% up to the saturation point of 1·7%, although not uniformly in any

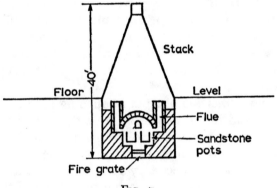

FIG. 7
CEMENTATION FURNACE

one bar, the carbon percentage decreasing from the outside of the bar to the centre. The product was known as 'blister' steel because the surface of the bar was covered with small blisters caused from the expansive force of the carbon monoxide formed by carbon reacting with the oxides of the slag mixed with the wrought iron. The blister steel bars were broken into short lengths, piled on top of each other and heated and welded together by hammering into larger bars. This was known as 'single shear' steel as it was largely used for making sheep shears and other cutting tools. If very high quality steel was desired the single shear steel bars were broken in half, and rewelded under the hammer into double shear steel, which was widely used for cutlery.

Crucible Process. Cement steel lacked uniformity and also

contained slag from the wrought iron used as the raw material. Benjamin Huntsman (1704–76), a Doncaster clockmaker who was dissatisfied with the quality of the blister steel and wanted a better material for his springs, became interested in the production of steel and conceived the idea of melting it in crucibles to achieve a homogenous product.

After many years of experimentation, Huntsman found that a mixture of certain fireclays from Derbyshire and Stourbridge, mixed in certain regulated proportions with china clay and coke dust, would stand up to the high temperature (1600°C.) necessary to melt the metal. The refractory fireclay pots, barrel-like in shape, varying in size from 12–18 in. in height with a capacity of 60–80 lb. of steel were charged with blister steel, charcoal and pure Swedish pig iron and covered with a lid. The pots were then immersed in a coke fire in a furnace holding up to 80 pots. Melting occupied 4–5 hours when the pots were lifted out of the furnace with a pair of tongs and the steel teemed into a mould.

The simple immersion of a pot in burning coke may seem an obvious step and in fact it was an idea that had been carried into practice many times before. Huntsman's achievement lay in the discovery of refractory materials capable of withstanding the high temperatures necessary to melt steel and capable of withstanding rough usage, e.g. being pulled out of the furnace with tongs.

Huntsman discovered the method in 1740 in Sheffield where he had settled. He attempted to keep the new process secret, but it was soon being operated elsewhere and by the mid-1900s some 1,500 crucible furnaces were in operation in Sheffield alone. From it stemmed Sheffield's reputation as a steel producer and until Bessemer invented his steelmaking process, Sheffield was the main centre of Britain's steelmaking, turning out some 60,000 tons of steel annually. The works which Huntsman built at Attercliff near Sheffield exist today, the firm being still noted for the high quality of its products.

Sir Henry Bessemer (1813–98). The Steel Age as we understand it was not to develop until the invention by Bessemer of his process in 1856. Bessemer was an engineer who among other things had made a fortune in inventing a new method for production of bronze powder (used in paint manufacture) which reduced the cost of production from £5 to 5s. per lb. He also gained large sums from his inventions in connection with glass, sugar and textile manufacture. In all he took out 117 British Patents between 1838 and 1883. He foresaw the great use to which steel could be put if it were cheaper than crucible steel (then some £60 per ton), and available in greater quantity. He also particularly wished to use it for ordnance, and incidentally this was one of the first uses to which it was put. His first experiments carried out at St. Pancras, in London, were with the fusion of pig iron and blister steel in a reverberatory furnace. In order to achieve a high enough temperature to melt the material, the air draught was increased by enlarging the grate area and by forcing jets of hot air through holes in the hearth side of the firebridge. This caused an intense flame which was sufficient to fuse the charge and achieve a high degree of decarburization.

At this stage Bessemer was approaching within measurable distance of the open-hearth furnace and might well have been led in time to the open-hearth process.

Bessemer notes that 'by chance some pieces of pig iron attracted my attention by remaining unmelted in the great heat of the furnace and I turned on more air through the firebridge with the intention of increasing the combustion. On again opening the furnace door after an interval of half an hour the pieces of pig iron still remained unfused. I then took an iron bar with the intention of pushing them into the bath when I discovered that they were merely thin shells of decarbonized iron, showing that atmospheric air alone was capable of wholly decarburizing grey pig iron and converting it into malleable iron without puddling or any other manipulation. Thus, a new direction was given to my thoughts, and after due deliberation I became convinced that if air could be brought into contact with a sufficiently extensive amount of molten crude iron, it would rapidly convert it into malleable iron'.*

In order to find out Bessemer designed his first converter— a fixed vertical vessel about 4 ft. high with six tuyères at the foot through which air was blown at 10–15 lb. pressure. Fireclay material was used as lining. About 7 cwt. of molten pig iron was run in; 'all was quiet for about ten minutes but soon after a rapid change took place, in fact the silicon had been

---

* Bessemer, Sir Henry. 'An Autobiography', *Engineering*, London, 1905.

quietly consumed and the oxygen next uniting with the carbon set up an ever-increasing stream of sparks in a voluminous white flame. Then followed a succession of mild explosions throwing molten slag and splashes of metal high up into the air, the apparatus becoming a veritable volcano in a state of active eruption. All this was a revelation to me as I had in no wise anticipated such violent results. However, in ten minutes more the eruption had ceased, the flame died down, the process was complete. On tapping the converter into a ladle and forming the metal into an ingot it was found to be wholly decarburized malleable iron.' Such were the conditions under which the first charge of pig iron was converted in a vessel neither internally nor externally heated by fire.

On August 11th, 1856, Bessemer presented a paper 'On the Manufacture of Malleable Iron and Steel without fuel' to the annual meeting of the British Association for the Advancement of Science at Cheltenham.

The paper aroused world-wide interest and many applications were secured for licences to manufacture malleable iron by the Bessemer process. The results, however, of its practical application by the licensees were 'most disastrous' to quote his own words, the metal poured from the converters being found to be brittle, breaking under the impact of a hammer. The cause of the failure lay in the high phosphorus content of the pig iron, for this element was contained in the majority of the ores both in this country and on the Continent. By purest chance, Bessemer himself in his original work had used a Blaenavon grey pig iron which was singularly free from phosphorus. Eventually, having traced the source of the trouble, Bessemer expended much time and money in vain attempts to overcome this handicap.

In the meantime another defect in the process had cropped up: on casting the metal, the resulting ingot was so full of blowholes as to be useless. Robert Mushet (1811–91), son of an ironmaster in the Forest of Dean, realized that the blowholes were caused by the evolution of carbon monoxide ($FeO + C = Fe + CO$) and that the remedy was to add a deoxidizer to combine with the excess of iron oxide; he suggested the use of spiegel (an alloy of manganese and iron) which would have a degasifying effect. As spiegel also contains carbon the material could be used to adjust the carbon content of the metal which was not accomplished in the Bessemer process as it then stood. After some confirmatory work Mushet in 1856 filed a patent for the use of spiegel for refining metal made by the Bessemer process.

The main problem, however, still remained unsolved and Bessemer was driven to the conclusion that his process was limited to phosphorus-free iron. After the initial failure, Bessemer's reputation had fallen so low that no ironmaster could be induced to give the process a further trial and so he resolved to set up his own steelworks. This he did in the middle of Sheffield, and in 1858 commenced operations, mainly making high-class tool and cutting steels which he sold at around £40 per ton, against the £50–£60 price of the Sheffield crucible steelmakers.

Thomas and Gilchrist. The phosphorus difficulty was eventually solved by Sidney Gilchrist Thomas (1850–85), a clerk at the Thames Police Court, Stepney. He was interested in chemistry and at a course of lectures at the Birkbeck Institution was made aware of the phosphorus problem when one of the lecturers said that 'the man who eliminated phosphorus in the Bessemer converter would one day make a fortune'. Having observed that Bessemer's converters were lined with siliceous material he conceived the idea that the phosphorus in the pig iron formed phosphoric acid on being oxidized by the air, the phosphoric acid so formed being reduced to phosphorus again by the siliceous (acid) lining and re-entering the metal. He realized that some other lining than a siliceous one would have to be used, which would combine with the phosphoric acid when set free. Thomas had a cousin, Percy Gilchrist, who was a chemist at the Blaenavon steelworks in South Wales, and with his co-operation (Thomas travelling down from London and spending the week-ends in Wales) a series of tests were carried out in 1877 using a small converter lined with lime bonded with sodium silicate. The results eventually demonstrated that with a basic lining and limestone additives phosphorus could be removed and poured off in the slag. The work was later transferred to the Middlesbrough works of Bolckow Vaughan & Co. a 30 cwt. converter being used, different limestones being tried, the final solution being a rammed lining of ground dolomite bonded with tar.

From this period onwards the Thomas Gilchrist basic process was widely used for the treatment of phosphoric ores, especially on the Continent due to the phosphoric nature of the ores of Alsace and Lorraine. It did not make the same progress in this country because by that time Siemens open-hearth steel had acquired a better reputation for quality than that of basic Bessemer steel; the Admiralty, for example, maintained that Bessemer steel was unreliable and demanded careful and delicate treatment in fabrication. Thomas died in 1885 from overwork when only in his thirties.

William Kelly. During the period in England when Bessemer was developing his process William Kelly (1811–88) was experimenting on the same lines at his ironworks in Kentucky. The wrought iron produced by Kelly was made by oxidation of the carbon in pig iron in presence of charcoal fuel. One day Kelly noticed that some molten pig iron in the bath not covered with charcoal became very much hotter when blown with air by virtue of being decarbonized. He realized (as Bessemer had done) that a blast of air on hot iron could act as a decarburizer, and deciding to act on this thought, built a small converter in 1851. He did not, however, apply for a patent until June 1857, but on learning of the American patent issued to Bessemer in 1856, he was able to prove that he had worked on the idea as early as 1847, and so Bessemer's patent was set aside in favour of Kelly's.

During the next ten years considerable litigation took place before the two men settled their differences and merged their interests. Because, however, of Bessemer's progressive leadership in the years to follow, the process became permanently identified with the name of Bessemer. The first steel made in America on a commercial basis by the pneumatic converter was produced at Wyandotte, Michigan, in 1862, by the Kelly Pneumatic Process Co. in a $2\frac{1}{2}$-ton converter, and by 1871 the annual Bessemer steel production in America had increased to 45,000 tons amounting to about 60% of the total steel production. A number of mechanical improvements were introduced in America, notably the detachable bottom, which cut down tuyère repair times considerably. Also larger converters were introduced, the U.S. Steel Corporation, in 1901, for example, operated 35 Bessemer converters ranging from 5–17 tons with a combined capacity of 7·5 million tons per annum, whilst later converters of 30 tons were operated with production rates of 40,000 tons per month. The number of tuyères was increased with the use of larger volumes of air. By 1880 American production of Bessemer steel exceeded that of the United Kingdom, totalling 1,047,000 tons against our 1,044,000.

MODERN PRACTICE

Bessemer. The present-day converter (Fig. 8) is an egg-shaped vessel 20–25 ft. in height, refractory-lined and supported on two

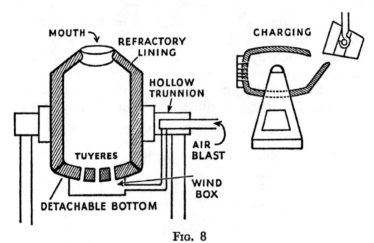

FIG. 8

BESSEMER STEEL CONVERTER

trunnions upon which it can rotate in a vertical plane for the purpose of charging and pouring. One of the trunnions is hollow to allow for the admittance of the air blast, the bottom being pierced for this purpose with 25–35 tuyères for the introduction of the air blast to the molten metal. The tuyères are subject to severe wear lasting only from 25–35 blows and hence the bottom is made detachable, so that a new one may be fitted without loss of time. This invention was due to A. L. Holley (1832–82), an American engineer, and made possible the replacement of a corroded bottom in 20 minutes or so. The air blast is generated at 25–30 lb. per sq. in. pressure by turbo-blowers, although in older shops steam-driven vertical recipro-cating-type blowing engines are still in use.

Bessemer converters may be either acid or basic-lined according to the impurities that it is necessary to remove from the blast-furnace iron. The former, which was the one Bessemer himself used, does not remove phosphorus and sulphur. The basic-lined converter does, however, and hence has a much wider application.

Basic Bessemer (or Thomas) Process. The lining used in this method is crushed dolomite mixed with tar and rammed into position whilst hot, using formers, and then dried by the use of a coke fire. Such a lining will last for approximately 150–200 blows.

In operation lime is first added to the converter to the amount of 200–300 lb. per ton of metal. Pig iron is then poured in, the converter turned up and the blow commences. Silicon and manganese are the first elements to be eliminated, their removal, however, not being accomplished by direct oxidation of the blast, but by iron oxide which is first formed.

$$2Fe + O_2 = FeO$$

this in turn reacting with the elements,

$$2FeO + Si = SiO_2 + 2Fe$$

$$FeO + Mn = MnO + Fe$$

the silica and manganese oxide combining with excess iron to form slag. During this period of the blow a short transparent flame appears at the mouth of the converter and lasts for about five minutes, when it increases in brilliance extending to as much as 30 ft. beyond the mouth of the converter. This indicates the removal of carbon by reaction with dissolved iron oxide.

$$FeO + C = Fe + CO$$

and also by oxygen of the blast

$$C + \tfrac{1}{2}O_2 = CO$$

The finish of the removal of carbon and commencement of elimination of phosphorus is marked by dense brown fumes in the flame due to the presence of oxides of iron and lasts about 3–5 minutes. Removal of phosphorus involves its oxidation to the pentoxide.

$$5FeO + 2Fe_3P = 11Fe + P_2O_5$$

the pentoxide then combining with the lime to form a basic lime phosphate slag, $4CaO.\ P_2O_5$. Phosphorus is the chief heat producer in the basic process and for this reason it is essential that the pig iron contains an adequate amount, $1\cdot5\%$ being regarded as a minimum. Only a small amount of the sulphur present in the pig iron passes into the slag, the majority remaining in the metal.

As practically all the carbon content of the metal has been eliminated during the blow, it is necessary to recarbonize, usually by addition of the appropriate amount of pig iron of known composition. Anthracite coal is also used for this purpose. First, however, the charge has to be slagged, the converter being turned down, and the blast turned off, the slag being poured into slag pots.

The vessel is then turned up, the blast restarted and the recarbonizers added, to effect the required composition. The steel is then poured into a ladle, deoxidation (killing) taking place by addition of ferrosilicon or aluminium. Total duration of the blow is about 15 minutes at a consumption of 12,000 cu. ft. of air per ton of steel produced. The phosphoric acid content of the slag produced approximates $15-20\%$ with a slag volume of about 20% of the metal. The presence of phosphorus renders the slag of value as a fertilizer, the sale thus favouring the economics of the process.

Acid Process. The lining of the converter used for the acid process is of silica brick or powdered ganister rammed to a depth of about 12–15 in. Life of this type of lining in the lower part of the converter is about 150 blows, that of the upper part being 300–400 blows.

Operation of the acid converter is similar to that of the basic process, the chief difference being that as the pig iron contains little or no phosphorus there is no 'afterblow', exhaustion of carbon in the bath being followed by subsidence of the flame indicating that the blow is at an end. The heat to keep the metal in the liquid state is supplied by oxidation of silicon present in the pig iron. Recarburization and deoxidation to rid the steel of surplus dissolved oxygen are carried out as before.

Hot-Metal Mixers. The original Bessemer plants were supplied with the necessary liquid pig iron by melting solid pigs in reverberatory furnaces or cupolas (page 299), a vertical shaft-type furnace, coke being used for fuel, air being blown in near the bottom.

The use of the cupola had certain inherent disadvantages, chief amongst them that the sulphur and phosphorus in the coke was absorbed by the iron, and as the acid process was incapable of removing these objectionable elements, the resulting steel was of poor quality. This led to the development of large refractory-lined gas-heated tanks in which the molten pig iron could be stored until required by the converters. Originally introduced in the United States by W. Jones of the Carnegie Steel Co. in the late 1880s, the first British mixer was installed at the Barrow melting shop in 1890. In addition to storage, variation in the composition of different blast-furnace irons can be evened out. With the introduction in the 1890s of the Saniter process of desulphurizing pig iron by the addition of manganese and other desulphurizing reagents, 'active mixers' came into use, the addition of reagents to the mixer causing a reduction in the amount of sulphur. This operation was later extended to the partial removal of silicon by addition of iron and limestone, these reagents forming a slag. The modern plant has usually two or three of these mixers which take the form of large cylindrical refractory-lined vessels, gas- or oil-fired, mounted horizontally and capable of rotation about their central axis for the purpose of charging and pouring off the hot metal as required. Capacity of these vessels varies from 200–1,500 tons. By mounting on a carriage the mixer is made transportable.

The Integrated Plant. The conservation of heat effected by the mixer is but one example of the heat economy effected in iron and steel plants. At every stage in the making and shaping of steel, the material must be maintained at a high temperature in order for a required operation to be effected.

Since fuel represents the largest single item of expense, fuel conservation is vital and has led to the integration of the entire process, blast furnace and coke oven, steelmaking furnace and rolling mill all being erected on one site. This not only makes possible the delivery of molten iron from the blast furnace to the steelmaking furnace and the transport of ingots to the rolling mills while still red hot; it also enables the steel works and rolling mills to utilize the blast furnace and coke-oven gases for which an economic outlet might not otherwise be available. With modern facilities and efficient practice in a well balanced and integrated plant, it is possible to produce one ton of ingots with one ton of coal.

Siemens Brothers. In the same year in which Bessemer took out a patent for the manufacture of steel in a converter, Frederick Siemens (1826–1904) a naturalized Englishman of German birth, filed a patent for the application of heat regeneration to furnaces. In conjunction with his brother, C. W. Siemens (1825–83) the first experimental furnace was built in 1858, a fireplace being built at either end of the furnace with checkerwork regenerators built underneath the furnace, the waste combustion gases passing through the regenerators, and the checker brick absorbing heat which was subsequently given up to the cold air required for combustion. Although economy in fuel was achieved and a high temperature attained, many difficulties had to be overcome before the principle of heat regeneration was achieved, not the least being the blocking of the checker chambers by the ash from the fuel. (Incidentally, during this period Siemens became associated with E. A. Cowper who successfully applied regeneration to the stoves which bear his name.) Eventually it was realized that these difficulties vanished if the solid fuel was gasified prior to burning it in the furnace. This gas also could then be preheated as well as the air. The next step was taken in 1861 when a patent was taken out in the names of both brothers for a 'gas producer'. The essential feature of this is that air is blown through a bed of incandescent coal or coke, the product of the reaction being carbon monoxide, the basic reaction being

$$2C + O_2 = 2CO + 4{,}400 \text{ B.T.U./lb. C.}$$

The efficiency of the process is improved by blowing steam into the producer, the red hot carbon decomposing steam into carbon monoxide and hydrogen.

$$C + H_2O = CO + H_2$$

The first regenerative furnace using gas as fuel was erected in 1861 at a glass works in Birmingham, England, and effected a saving in fuel of 50%. By the end of 1862 about 100 gas-fired furnaces were in operation. The furnace was first used in the ferrous industry for making crucible steel, and although during the next few years attempts were made at several British works to refine steel, indifferent results were obtained. Siemens was thus forced, as was Bessemer before him, to erect a furnace of his own, which he did in Birmingham and which at first was employed in remelting crucible steel. A year later in 1867 he took out his first patent for making steel, the patent specifying that cast steel may be produced by melting pig iron in the furnace, the superfluity of carbon being eliminated by addition of iron ore which reacted with the carbon, silicon and manganese forming a slag, spiegel then being added for recarburizing and degasifying the steel. This method subsequently became known as 'the pig and ore' process, as distinct from the 'pig and scrap' process which was originated by Pierre Martin and his father Emile at Sireuil in France. The Martins had taken out a licence from Siemens in 1863 and had erected a regenerative furnace in which pig iron was melted, malleable iron then being added in suitable proportions to reduce the carbon content to the figure required. By the substitution of steel scrap for the malleable iron (or the ore in Siemens pig and ore process), the Martins found it possible to dilute the charge to such an extent that less oxidation was necessary. The scrap method was named the Siemens–Martin process and is today the commonly used method of making open-hearth steel, for it enables the large quantities of steel (and iron) scrap which become available every year to be utilized as a raw material for steelmaking.

The term 'open hearth' probably originated from the fact that the metal was melted on the bed of a furnace, open as opposed to the closed or crucible method. The advantages of the process may be summed up as follows:

(1) Owing to the fact that the elimination of the impurities takes place gradually (from 6–12 hours compared to the 20 minute Bessemer blow), the composition of the bath is under much better control than in the Bessemer process.

(2) Scrap metal in practically any proportion can be utilized as a raw material, whereas in the Bessemer process it is limited to about 5%.

(3) The basic operation process does not depend on phosphorus for its successful operation as does the basic Bessemer process, which demands 1·7–2·0% phosphorus in order to maintain a temperature high enough for the reaction to proceed.

(4) Steels made by the open-hearth method contain much less nitrogen than those manufactured by the Bessemer process (nitrogen being absorbed from the air blow, rendering the metal liable to embrittlement.)

The first large company formed especially to operate the Siemens process was the Steel Company of Scotland, established in 1871; and in Boston, Massachusets, a 5 ton acid-lined furnace operated about the same time. Since these early days furnace design has suffered no radical change. Up to the 1880s the furnaces were charged manually through a charging door by means of long-handled shovels which rested on a fulcrum bar placed across the furnace opening. The charge consisted of a mixture of pig iron and scrap iron and was run down in about 8 hours and then issued through a tapping hole and teemed into moulds. (Hot metal in place of cold pig iron was first used in the open-hearth furnace in the late 1880s.) The charging involved hard, hot labour and in America a charging machine was invented in 1887 by S. T. Wellman, the first machine being installed at the Lakeside plant of the Otis Steel Co., Cleveland. Early in 1886, one of the 15-ton Otis furnaces was changed from an acid bottom to a basic bottom with magnesite imported from Austria. This is reported to be the first use of a basic open-hearth furnace in the U.S. In 1895 Wellman invented another piece of equipment, the electric magnet which loaded pig and scrap iron into charging pans which are picked up by the charging machine and emptied into the furnace. These two inventions reduced the number of men employed in charging by at least half, in addition to eliminating much hard labour. Charging times were decreased, delays reduced, and the output in tons of steel per ton was greatly increased. Succeeding furnace developments were concerned with improvements in operation aimed at increasing production, improving quality and reducing costs. Of these advances only a few can be listed here.

1. Improved furnace refractories resulting in increased furnace life.

2. With the use of producer gas as fuel, roof life was low, the furnace brickwork tended to be overheated, extensive checker-work constructions were necessary to preheat the gas, and gas ports were difficult to maintain in good repair. With the change over to fuel oil in the late 1930s, these disadvantages were largely eliminated, for this type of fuel was cleaner, more dependable, gave a higher flame temperature and led to higher rates of furnace production.

3. Improved instrumentation and furnace controls.

4. Water cooling of ports, doors and frames. Before 1915, charging and working areas in front of the furnace were covered with steel plates. These got hot and often warped by the intense heat, making working conditions on the charging floor almost unbearable.

5. A new phenomenon is the increasing use of oxygen to melt down scrap iron, and in the later stages to reduce the carbon content of the molten metal. Such a procedure increases tons per hour and improves the quality of the steel.

OPEN-HEARTH PROCESS

Pig iron together with steel scrap and a flux are charged to the furnace in regulated quantities, order and compositions. Oxidation of impurities induced by the furnace gases and addition of iron ore proceeds by the same chemical reactions as occur in the Bessemer process, but at a much slower rate. As with the Bessemer process, both acid and basic processes are operated, but as the basic open-hearth furnace removes phosphorus and sulphur, there is a much wider choice of raw materials; at present about 85% of the steel produced comes from basic furnaces.

The Furnace. The open-hearth furnace (Fig. 9) is of the reverberatory type, essentially a shallow hearth with side and end walls surmounted by an arched roof, fuel being supplied at each end of the furnace.

Hearth. The hearth consists of a refractory-lined steel pan supported on steel beams, the dimensions for a 100-ton furnace being approximately 450 sq. ft. and 30 in. in depth. The basic hearth is prepared by adding a little water to powdered magnesite, mixing and shovelling on to the hearth. It is then rammed into position with pneumatic hammers, dried and fired at above 1600° C. for 24 hours. The life is about 200 heats.

Under the hearth are two pairs of checkerwork regenerators, one pair absorbing heat from the waste furnace gases, while the other gives up heat to the gas and air required for combustion. At regular intervals the direction of the gas and air is reversed,

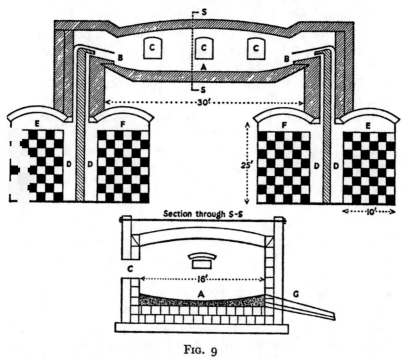

FIG. 9

OPEN-HEARTH FURNACE, showing (A) Hearth (B) Fuel ports (C) Charging doors (D) Slag pockets (E & F) Regenerators (G) Tapping spout

so that each pair of regenerators alternate in absorbing heat from the waste gases, yielding it to the incoming air and gas. High-duty fireclay brick is commonly used for the checkerwork. One of the worst nuisances is dust from the waste gas which necessitates frequent recleaning. When fuel oil is used, only one pair of regenerators is required, for only the air is preheated.

The roof arches over the hearth from the front to the back of the furnace. It has to withstand the corrosive effect of dust and fume at extremely high temperatures and must possess sufficient strength and rigidity to maintain the arch stress, for it is self-supporting over the whole span of the hearth. Silica brick is commonly used, for it possesses good load-bearing capacity at high temperatures, constancy of volume, and a high temperature of incipient fusion. The recent introduction of oxygen lancing to open-hearth steelworks has, however, necessitated a change to chrome-magnesite brick as roof lives were extremely short with silica brick.

Side and end walls are generally composed of chrome magnesite brick for they are exposed to high temperatures, the splashing of metal and slag and in the case of the end walls erosive action of the dust particles in the rapid stream of moving gases. Charging doors numbering 3–5 are located in the front wall, a tap hole for emptying the furnace being in the back wall.

Ports at either end of the furnace serve to admit fuel and air and provide for escape of the waste combustion gases. Fuel is either fuel oil, producer gas or coke-oven gas, or a mixture of blast-furnace gas and coke-oven gas.

OPERATION

A typical 50–50 charge consists of 50% steel scrap and 50% molten pig iron. The steel scrap, limestone flux (10% of the weight of the charge) and iron ore are first charged by the aid of a charging machine through the doors in the front wall of the furnace. When charging is complete, which takes 2–4 hours, the heat is turned on. Oxidation of the scrap begins and about 2 hours after charging, when the temperature has attained 1100–1200° C., addition of molten pig iron commences. Following this addition vigorous action takes place, the silicon, manganese, phosphorus and carbon of the pig iron being oxidized by the ferrous oxide present in the bath.

$$FeSi + 2FeO = SiO_2 + 3Fe$$

$$Mn + FeO = MnO + Fe$$

$$2Fe_3P + 5FeO = P_2O_5 + 11Fe$$

$$Fe_3C + FeO = CO + 4Fe$$

The reactions are all exothermic and thus help in keeping the bath molten during this period. The SiO_2 and MnO form slag whilst the oxidation of carbon to carbon monoxide gas and its escape from the bath causes considerable foaming, being technically known as the 'ore boil'. This is advantageous in so far as the stirring action promotes uniformity of temperature throughout the bath. As the carbon content of the bath decreases and the temperature rises, the ore boil subsides and calcination of limestone commences.

As it progresses carbon dioxide is given off

$$(CaCO_3 = CaO + CO_2)$$

causing turbulence of the bath, known as the 'lime boil'. Solid lime rising to the surface of the bath serves to remove the phosphoric acid and some of the sulphur as a basic slag. After the lime boil has subsided, the refining period begins, the object being to reduce the phosphorus, sulphur and carbon to the required levels and to permit the finishing off of the operation. Samples of the metal are withdrawn from the bath from time to time, and the carbon content determined. As it approaches the desired level, the feeding of ore is stopped and 'blocking of the heat' is accomplished by addition of deoxidising agents which combine with the dissolved oxygen in the metal and thus stop or block any further oxidation of the carbon. When the heat of steel is ready to be tapped, the clay-plugged tap-hole is opened up, and the steel run down the tapping spout into a ladle, any alloying or recarburizing materials being added to the spout or ladle. The steel is followed by the slag which overflows the ladle into slag pots. The ladle is then lifted by an overhead crane, and conveyed to the ingot moulds. The total time consumed during the various stages of a heat varies from 8 to 12 hours depending on the nature of the charge and the final carbon content desired.

For a 100-ton 50:50 charge the duration of operation is about as follows:

	hours
Charging	2
Melt down	3
Hot metal addition	$\frac{1}{2}$
Refining	4
Repairs to furnaces	$\frac{1}{2}$
	10

Acid Open-Hearth Process. The acid process is in general limited to the production of steel castings and ingots for forging. It does not permit the removal of phosphorus and sulphur, hence its use is confined to raw materials containing less than the specified amount of phosphorus and sulphur required in the finished steel. Because of the care that has to be exercised in the choice of raw materials the steel produced in an acid furnace is, however, generally regarded as superior to that produced in a basic furnace.

The furnace in its general features is similar in construction to that of the basic furnace except that the hearth is of siliceous material. Silica brick to a depth of 12–20 in. is built up on the pan, successive layers of sand being fritted into position at a temperature of about 1600° C. until a depth of 12–18 in. is attained.

The raw materials consist normally of cold pig iron and specially selected scrap. During melting the silicon and manganese will have been partially oxidized by the furnace gases and a slag will have formed. Iron ore is now added which oxidizes the remaining silicon and manganese, oxidation of the carbon giving rise to the ore boil. On the cesssation of the boil a little limestone is added to thin the slag, and as the carbon approaches the required level, further carbon drop is blocked by addition of ferrosilicon or ferromanganese. The metal is then tapped, any finishing additions being added to the ladle.

ELECTRIC STEELMAKING

The use of electricity as a source of heat for melting steel takes two forms. An electric arc may be struck between electrodes which project through the roof of a suitably shaped furnace in which is contained the raw steelmaking materials, the heating of the material taking place by radiation from the arcs, and by the resistance of the metal to the passage of the current.

In the second method a current is induced within the metal, the resistance of the metal to the flow of current resulting in the production of heat. This method is known as induction heating, and exercises a melting rather than a refining action; hence the furnace is used mainly for the production of high grade alloy steels from selected scrap material.

The Arc Furnace. Sir William Siemens in 1878 was the first to employ the electric arc for smelting metal in a closed hearth. The arc furnace which he devised was demonstrated in 1879 to the members of the Iron and Steel Institute at Siemens Brothers Charlton Works, when 5 lb. of steel scrap was melted in about 20 minutes. At this early date electric power was limited and expensive and the development of the electric melting-furnace had to await the expansion of the electrical industry. Various electric arc furnaces were constructed, mostly on the Continent, and in 1900, Dr. Paul Heroult (1863–1914) produced the first successful commercial direct arc steelmaking furnace at La Praz, France. The Heroult patent covered single or multi-phase furnaces, with the arc in series through the metal bath, the three-phase type of furnace having proved to be the most successful in the production of steel. Another possible use which had been envisaged by Siemens was the electric reduction of iron ores, dispensing with the blast furnace; and this has proved of interest to countries with inadequate fuel resources but with large water resources for hydro-electric generation such as Sweden and Norway.

While the use of the electric furnace in Britain was retarded by the cost of electric power, use was made of it in Germany for a process known as duplexing, which is a term applied to a combination of processes for manufacturing steel. As carried out in Germany and elsewhere, the method consisted of blowing molten pig iron in the Bessemer basic converter until the silicon manganese and most of the carbon had been oxidized, and then transferring this semi-finished metal to a basic-lined arc furnace where the phosphorus and the remainder of the carbon are oxidized to the desired limit. The method avoids the high cost of smelting cold charges by electricity, and due to the better control possible in an electric furnace provides uniformity from heat to heat. Duplexing, using the open-hearth furnace in place of the electric furnace, has also been employed in many countries.

American steelmakers, quick to sense the possibilities of electric smelting, invited Heroult to visit their country and in 1905 the first American arc furnace for steelmaking was installed in the plant of the Halcomb Steel Company of Syracuse, N.Y. This was a single-phase two-electrode furnace powered by a 500 kW generator producing high amperage at low voltage. The furnace was rectangular in shape, of 4-ton capacity, the product being tool steel. Three-phase cylindrical furnaces were developed about 1910 as was also top-charging, which replaced hand-charging, the roof of the furnace being lifted and swung aside, the entire charge then being placed in the furnace by drop-bottom buckets. With the impetus thus afforded by these and other inventions the trend towards larger furnaces was initiated. From the early ratings of 500 kW units and voltages of 90 V and capacities of 4 tons, ratings increased to 7,500 kVa and 274 V with capacities of 15–20 tons by the 1920s, with automatic regulators for balancing current and voltage, in order to maintain the desired length of arc. As capacities became larger a tilting type of furnace was evolved for convenience in tapping and slagging, the furnace being mounted on toothed rockers which rested on toothed rails, the tilting being controlled by a motor-driven rack-and-pinion mechanism. Electrodes of amorphous carbon were being replaced by the denser and more conductive graphite with its better electrical characteristics. Water cooling was also introduced to effect improvement in refractory life.

During and after World War Two, many large basic electric furnaces, some up to 200 tons in capacity, have been installed and it is certain that this trend will continue, furnaces of 300-ton capacity having six electrodes and transformers with large capacities being possible. With its inherent advantages of flexibility and low repair costs the electric arc furnace has an impressive future, for where power costs are low the operating cost can approach that of the open hearth. Alloy and plain carbon steels of high quality can be made and the ferro-alloy industry has already for many years been based on its use.

The modern electric furnace is cylindrical in shape consisting of a steel shell lined with refractory material generally basic in character. The domed roof is removable, being capable of being slewed to one side for the purpose of charging with raw material from a drop-bucket skip. The roof is pierced by three circular openings spaced at the corners of an equilateral triangle, through which pass the electrodes, water-cooled rings around the electrodes acting as a seal.

Electrodes are made either from carbon or graphite, which although chemically similar differ widely in electrical and physical properties. According to the electrical rating of the furnace, electrodes are made in many sizes ranging from 6 in. in diameter for a transformer capacity of 2,000 kVa up to 20–25 in. diameter for 35,000 kVa. Under the action of the arc, the electrodes are gradually consumed at a rate of 15–25 lb. per ton—for graphite electrodes, carbon being consumed twice as fast—chiefly by oxidation, and accordingly means have to be provided for renewal. For this purpose a threaded nipple is inserted in the top of the old electrode and a new one is screwed down on the exposed end of the nipple.

Data relevant to a 100-ton Heroult furnace are given below:

Inside diameter	20 ft.
Inside height	7 ft.
Hearth thickness	3 ft. 9 in.
Rating, kVa	20,000/25,000
Electrodes (number)	3
Electrodes (diameter)	16 in.
Roof life in heats	120
Average heat time (tap to tap)	5 hours
Tons per hour	18·5
kWh per ton	500

Current Regulation. During the preliminary melting down of the solid charge much more power is expended than in the succeeding stages when the molten condition has been attained. This factor leads to wide variation in current requirements, further aggravated by the oxidation and spalling to which the electrodes are subject and which continually shortens their length and tends to produce a longer arc. As constant energy input is necessary for both economy and efficient operation, it is customary to include in the current circuit automatic-current regulators. The basic principle used to cope with current irregularity is that as the length of the arc governs power input, any increase or decrease in its length induces a corresponding increase or decrease in voltage. Hence by lowering (or raising) the electrodes, fluctuations in the length of the arc can be evened out. In one method (known as the rotary regulator) control over the variations is exercised by the raising and lowering of the electrode by electric motors, operated automatically by variable voltage controls which balance currents in the electrode circuit against the voltage across the arc. In this manner each electrode motor maintains in its electrode the desired current. Raising or lowering of the electrodes is accomplished through wire cables, each separately connected to a winch and the control motor. Electronic devices are now available which automatically rectify any out-of-balance variation between voltage and current, by applying the electrode motors with the correct power. In another development the power control unit is linked to a switch operated by a thermostat placed in the roof of the furnace, control thus being operated by temperature.

In operation the electrodes are lifted, the roof slewed out of the way and the material charged via the drop-bottom skips. When charging has been completed the roof is replaced, and the electrodes are lowered to about an inch above the material.

1. *Melting down period.*   The material immediately below the electrodes melts first and forms a pool on the furnace hearth, the electrodes activated by the automatic control following the charge as it melts down. This melting down operation which is the most expensive period as regards power consumption occupies about 2 hours.

2. *Oxidation period.*   The oxygen necessary for oxidation of the metalloids present in the metal was supplied by iron ore or mill scale, but recently this has been superseded by the introduction of gaseous oxygen to the molten bath via a steel pipe or lance at a rate of about 500 cu. ft. of oxygen per minute.

The use of oxygen possesses many important advantages among which is the rapid removal of carbon from the bath, normally a slow operation; and further, since the reaction is exothermic, electricity supply may be shut off for a period and thus power consumption can be saved.

3. *Reduction period.*   The slag formed by the oxidized products is removed from the surface of the bath by shutting off the power, raising the electrodes, back-tilting the furnace and then raking the slag out through a door into a ladle. It is replaced by a strongly reducing slag composed of lime, fluorspar, sand and powdered coke or anthracite, and serves for the purpose of reducing the oxides in the bath and also for removal of sulphur

(as calcium sulphide). At the high temperature of the arc, lime and carbon react to form calcium carbide,

$$CaO + 3C = CaC_2 + CO$$

the sand and fluorspar serving to flux the lime. The carbide reduces any oxides present in the bath

$$CaC_2 + 3MnO = CaO + 2CO + 3Mn$$

$$CaC_2 + 3FeO = CaO + 2CO + 3Fe$$

desulphurization taking place as follows:

$$2CaO + CaC_2 + 3FeS = 3CaS + 2CO + 3Fe$$

the desulphurization being aided by high temperature and high manganese content. Ferrosilicon and ferromanganese are then added to complete the deoxidation, the power is shut off, the electrodes raised, the furnace tilted forward and the metal poured into a ladle.

In addition to carbon steels the production of steels alloyed with high percentages of oxidizable metals can be produced under reducing conditions much more efficiently in the basic electric furnace than elsewhere.

Ferro-alloys also are commonly produced in the electric arc furnace, which is essential for those requiring a high temperature for their formation, and for very low carbon alloys.

Induction Furnaces.   When a conductor passing an alternating current is placed in close proximity to another conductor, lines of electro-magnetic force radiate out and produce eddy currents in the second conductor and are dissipated as heat. In its application to the melting of metals an alternating current of high frequency is passed through a copper coil, inside which is a crucible containing the metal to be melted. The magnetic flux created by the high-frequency current passes through the metal charge, the induced current melting the charge by the heat developed by the resistance of the charge to the current.

The first attempt to apply this principle was made by S. Z. de Ferranti in Italy in 1877. This was a high-frequency furnace consisting of a U-shaped refractory vessel, the metal contents being melted by means of a primary external coil. It, however, had no commercial application until Kjellin installed and operated one in Sweden in 1899, and in 1907 an experimental furnace in Sheffield produced 2-ton steel castings. The first large installation of this type was made in 1914 at the plant of the American Iron and Steel Co., Pennsylvania, but was not successful.

These early furnaces were operated with very high-frequency oscillating current supplied by static converters of the spark gap type. The kW input was restricted with this type of converter, but it allowed the melting of small charges to be accomplished. The invention of the motor generator set in 1925 with frequencies in the range 500–3,000 made possible the industrial expansion of the high-frequency furnace.

The researches of Dr. E. F. Northrup at Princeton University, U.S.A., in 1922 were responsible for the development of the high-frequency induction furnace, and the first furnace of this type for the production of steel on a commercial scale was installed at Sheffield in 1927, and in Pittsburgh a year later. These furnaces had a capacity of 600 lb. of steel, and were served by a current of frequency 960 cycles per second.

In a sense the induction furnace can be regarded as a development of the crucible process, the former gradually replacing the latter in the production of high quality tool steels. As with the crucible process, no refining action takes place in the induction furnace; what goes in comes out, which means that pure raw materials have to be used, their composition adjusted to that required in the finished melt.

The type of furnace in use today is crucible-shaped and consists of powdered ganister or quartzite mixed with a binder and rammed into position around a steel former shaped to the required outline of the crucible, the former being subsequently melted away with the first charge of steel. Water-cooled copper tubing surrounds the shell of the crucible and serves as the conductor for the primary high-frequency current. For convenience in pouring, the furnace is mounted on trunnions capable of forward tilting by an electrical mechanism.

An interesting feature of the process is that the heat is developed mainly in the outer rim of the metal in the charge, and carried to the centre of the metal by conduction. Electromagnetic action exerts a strong stirring action (motor effect) on the metal causing it to circulate, giving the surface of the bath a convex shape varying in intensity with the power input. This stirring effect is of advantage in giving uniformity to the melt.

The furnace requires about 600–700 kWh of input power per ton of molten metal, heats being melted in about 2–4 hours, according to the capacity of the furnace, which varies up to 10 tons.

Frequencies in the neighbourhood of 1,000 cycles are usual, being supplied either by rotary motor generation sets or mercury arc frequency-changers.

VACUUM MELTING

Production of steel involves a refining treatment in which sulphur and phosphorus in pig iron are eliminated as far as possible. Small quantities of foreign substances such as dissolved gases, non-metallic inclusions, etc., remain, and for certain engineering needs, metal with high reliability and purity is required. To meet these needs vacuum melting techniques have been devised whereby improved physical and mechanical properties are obtained, unattainable in any other way. The use of vacuum removes dissolved gases and volatile impurities from the melt, reduces metal oxides by the carbon-oxygen reaction and eliminates the effects of atmospheric contamination, thereby improving impact strength, tensile strength, ductility and fatigue strength.

While the vacuum melting of non-ferrous metals and alloys has been practised in Germany since the 1920s, notably by the Heraeus Vacuum-Schmelze, the tonnage melting of steels is a development only during the past ten or twelve years, prompted largely by the demand for high quality material for the gas turbine engine. Before World War Two, vacuum pumps and vacuum technology in general were not sufficiently advanced to maintain a high vacuum during the process of melting, but with the demand for materials which would withstand very high temperatures substantial improvements took place, vacuum pumping equipment capable of high capacity and pump speeds at low pressure (10^{-2}–10^{-3} mm.) being developed. Even now, only a beginning has been made in a field which promises to make a major contribution to the melting and casting of metals and alloys with improved properties.

Apart from the obvious advantages of eliminating an oxidizing atmosphere, the avoidance of gas porosity in the vacuum-cast ingot ensures that surface defects will not occur during machining, no forging treatment being required to weld gas cavities. Further reduced piping in vacuum-cast ingots leads to lower crop rejects, and the ingot may be fabricated or drawn into wire with better surface finish than can be obtained with the normal product.

Two main types of furnace are used for vacuum melting:

1. Induction furnaces of the high-frequency type.
2. Arc furnaces using a consumable electrode.

The main feature distinguishing the two types of furnace is that in the latter a direct arc is used as the heat source and there is no refractory crucible, an ingot of metal being built up by progressive solidification in a water-cooled copper mould.

Induction Vacuum Melting. This method was the result of investigation over thirty years ago by the Heraeus Vacuum-Schmelze, but lack of adequate equipment and vacuum technique hampered the application until after World War Two.

The elements of the modern vacuum induction furnace comprise a refractory crucible (surrounded by a high-frequency coil) to hold the charge and a mould contained within a welded steel cylindrical-shaped vacuum chamber.

High-capacity oil-vapour booster pumps provide the vacuum at a pressure of less than 10^{-3} mm. of mercury. In operation the raw material is charged to the crucible and the power applied. When molten the vacuum chamber lid is placed in position and evacuation begins. Once molten it is only necessary to supply power to compensate for heat losses and to provide adequate (natural) stirring of the melt to promote degasification and solution of the alloy additives. Special consideration is given to the problem of charging make up and alloy additions. In one method the charging system employs a number of buckets which travel in the inner periphery of the vacuum chamber and are constructed with drop-away flaps released by the action of a transfer chute. The operator is informed which bucket is in the charging position by an indicator on a selecting mechanism. A simpler procedure involves a vacuum lock feeding arrangement, alloys being charged into an external transfer chute located directly over the crucible without breaking the vacuum in the main chamber. This device can also be utilized for refeeding the whole of an initial charge without breaking the vacuum, thereby increasing the rate of production. On completion of the melt, the induction furnace is tilted, the metal flowing into an ingot mould, alternatively a fusible plug in the bottom of the crucible is removed, the metal pouring directly into a mould below the crucible. At the end of the casting operation the vacuum is broken, the mould(s) removed and replaced by fresh ones, and the process repeated. Provision can also be made for the mould chamber to be isolated by a vacuum lock from the main chamber thus permitting removal of the ingots without breaking the vacuum in the main chamber, thereby making provision for completely continuous operation. Capacities are normally of the order of 600 lb. or so, but furnaces in excess of 1 ton are in operation in the United States.

Vacuum Arc Melting. This form of vacuum melting employs a consumable electrode of the metal to be melted connected externally to the negative pole of a d.c. generator. A hearth is formed by a water-cooled copper tube and is connected to the positive pole of the generator. An arc passes from the electrode to the crucible and provides the heat for melting the raw material which is fed at a controlled rate into the arc by an electric motor. Provision is made for evacuation of the furnace.

The process was initiated in 1905 by W. Von Bolton of Siemens and Halske A.G. in Berlin who with the very primitive apparatus then available experimented with the high melting point (3000° C.) metal tantalum and succeeded in melting and obtaining compact ingots of the metal. W. J. Kroll in America in 1940 improved the Von Bolton technique and applied it to the melting of the highly reactive metal titanium, using a non-consumable type of electrode (tungsten). This was replaced in 1954 by a consumable type of electrode and the method applied to several other reactive and refractory metals such as zirconium, molybdenum and niobium.

In recent years it has been applied to the melting of heat-resisting steels, stainless steels for atomic energy purposes and ball-race steel yielding a product free from harmful non-metallic inclusions and gases. Using this technique steel electrodes are being arc-melted and cast under vacuum to produce 60 in. ingots weighing 40 tons. Further details of the consumable arc vacuum melting process will be found on page 245.

USE OF OXYGEN

In the basic converter process the oxidation necessary to remove impurities from the pig iron is done by injection of air. Apart from the generally lower reliability and quality of the steel, the process has two further serious limitations. Only pig iron with a phosphorus content between $1\cdot5$ and $2\cdot0\%$ can be used; and the resulting steel always contains a proportion of nitrogen which adversely affects ductility. These difficulties have their origin in blowing with air. Part of the nitrogen in the air blast is absorbed by the liquid metal, and the nitrogen which is not absorbed carries away so much heat in the gases that only metal of high phosphorus content will generate sufficient exothermic heat for the operation to be carried out. The remedy is to omit the nitrogen, replacing it with an oxygen blast. In addition to better quality steel the use of oxygen as opposed to air also confers better heat transfer and yields a higher reaction temperature and an increased rate of reaction.

First developments soon after the war consisted of enrichment of the air blast to the converter with oxygen. It was found that pure oxygen could not be blown through the metal owing to rapid corrosion of the tuyères. Although steels made by partial replacement of the air blast by oxygen were lower in nitrogen than standard Thomas steel they were not completely satisfactory for deep-drawn components such as car bodies. Following this the air blast was completely replaced by a mixture of pure oxygen and steam or carbon dioxide in roughly equal proportions. Carbon dioxide being comparatively costly is but rarely used. The use of the oxygen/steam process developed by the Belgian and German industry is now well established in Great Britain, the new Bessemer plant installed at the Abbey Works of the Steel Company of Wales in 1958 being entirely oxygen/steam blown for the making of V.L.N. (very low nitrogen) steels. The other two operators of the basic Bessemer process in this country, Richard Thomas & Baldwins, Ebbw Vale, and Stewart & Lloyds, Corby, have also introduced oxygen/steam or oxygen enriched air-blowing. In the past ten years four further oxygen processes have been developed in Europe, the L–D and LD/AC process in Austria, the Kaldo in Sweden and the Rotor in Germany. These processes have only been made possible because the manufacture of gaseous oxygen in bulk quantities has resulted in a much lower cost (2/- per 1,000 cu. ft.) than the normal liquid oxygen delivered in cylinders.

Linz–Donawitz (L–D) process. This method developed in Austria by the Vereingte Oesterreichisch Eisen und Stahlwerke of Linz and Oesterreichisch Alpine of Donawitz is now leading the field—some 30 million tons of capacity now (1962) being projected. In this process the converter is provided with a solid bottom, the bottom blowing being replaced by top blowing with a single water-cooled copper lance located at a distance of about 4 ft. above the metal surface. Oxygen at high speed issues from the lance and is directed on the surface of the liquid pig iron, penetrating the metal with production of an intense bright flame which continues at a uniform intensity until the carbon content of the metal is reduced to a level of about 0·05%. The converter is then slagged off and the steel poured into ladles in the normal manner. A large converter of 100 tons capacity requires 20–25 minutes on blast for a cycle time of about 50 minutes. The process is being introduced into Britain on a substantial scale and by 1965 production capacity should be some 4 million tons.

The process combines the advantages of low capital expenditure and high speed of working, yielding a steel with a very low nitrogen content. In addition it can handle much more scrap as compared to the conventional Bessemer process. Refractory consumption is very low, figures of less than 10 lb. per ton of steel being quoted at Linz for magnesite linings with lives of up to 600 heats.

LD/AC Process. The high temperatures attained in the L–D process produces a very fluid slag which whilst encouraging carbon removal does not effect dephosphorization, limiting the method to pig iron with less than 0·4% phosphorus. By the introduction, however, of powdered lime through the lance into the oxygen stream, elimination of phosphorus is effected, making possible the treatment of high phosphorus irons. The OLP (Oxygène Lance Poudre) process was developed by the French metallurgical research organization IRSID whilst the OCP (Oxygène Chaux Pulvérisée) process was developed by the Belgium metallurgical research body, CNRM, and brought to commercial success at the ARBED—Dudelange works at Luxembourg. The OCP method has since been styled LD/AC (i.e. after ARBED and CNRM) process. The difference between the L–D and LD/AC process is not fundamentally one of physical equipment but rather of technique. In operation the converter is charged with molten iron in the usual manner together with steel scrap and lime, after which the vessel is brought to a vertical position. The lance is lowered and the jet applied using oxygen alone for the first few minutes, after which the finely powdered lime is injected through the lance. After about 15 minutes the converter is turned down and the slag poured off. This slag will be low in iron and rich in phosphorus ($10–15\%$ P_2O_5). The converter is then turned up and the oxygen lime blow continued for a few minutes more when the steel is tapped, the second slag being retained in the vessel for use in the next heat, the high iron content of this slag increasing the yield of steel.

Kaldo Process. The word Kaldo commemorates both the inventor and the birthplace of the process, the inventor being Professor Kalling of Stockholm and the birthplace Domnarvets in Sweden, where the first commercial vessel of 30-ton capacity went into operation in May 1956. The Kaldo vessel is similar in shape to the L–D but the formation of a fluid slag is assisted by revolution of the furnace at speeds of up to 30 r.p.m. on an axis inclined at $17°$ to the horizontal. Variations in the rate of oxygen supply and speed of rotation enabled a close control of steel composition to be achieved. Carbon monoxide formed by oxidation of carbon in the iron is burnt to carbon dioxide within the furnace, the significance of this being that the formation of CO_2 releases approximately the same amount of heat as that due to oxidation of the impurities contained in the pig iron, the additional heat thus released being available for melting down large quantities of scrap steel.

Rotor. The Rotor process, developed by Dr. Graeff in Oberhausen, Germany, is somewhat similar to the Kaldo except that it is carried out in a horizontal vessel. The vessel revolves every two minutes, the object being to protect the refractories by cooling them continuously in the metal and slag. Oxygen is introduced through water-cooled nozzles both below and above the bath surface, the former performing the oxidation, the latter serving for conversion of the monoxide to dioxide with resultant high thermal efficiency.

Oxygen and Open-Hearth Furnaces. Two of the main disadvantages of the open hearth are the slow speed of working and the high fuel consumption. By the application of oxygen to existing open-hearth furnaces substantial improvements in both these factors can be attained. Oxygen-assisted combustion increases the intensity of the flame, thereby reducing the time required for melting down the charge. It has been established (by the Consett Iron Co. Ltd.) that the rate of production can be increased by about 35% if roughly 25% of the air is replaced by oxygen. Even bigger improvements have resulted from lancing of the steel bath to speed up metalloid removal. The Steel Company of Wales first developed this practice in 1953, water-cooled oxygen lances being inserted through the furnace roof to produce very low carbon steel essential for the highest deep-drawing quality. Many other steel companies throughout the world are now using the process, the Steel Company of Canada reporting outputs of 100 tons per hour in their 500 ton open-hearth furnaces, a figure of 4 or 5 times greater than without lancing.

The Ajax process used at the Appleby–Frodingham Steel Works is somewhat similar in operation, oxygen being injected via 30 ft. long water-cooled lances introduced into the furnace through the end wall in the position normally occupied by fuel burners, output thereby being nearly doubled.

ALLOY STEELS

The deliberate addition of certain elements to steel to obtain enhanced properties began in 1819 when Michael Faraday investigated the properties of alloys of iron with a large number of other elements, including chromium. He did not, however, pursue the matter, being deflected by his research into electromagnetic induction. He left records of his work and a large number of specimens which were assayed in 1931 by Sir Robert

Hadfield, who pointed out that had Faraday continued his investigations the Alloy Steel Age would probably have started fifty years earlier.

Addition of chromium to iron increases the wear resistance, and the United States were the first in the early 1870s to develop application of these iron–chromium alloys. In France, Brustlein at Unieux produced chrome pig iron and chrome steels for commercial use in 1877, the manufacture of blast furnace ferrochrome being undertaken soon after. In England John Brown & Company at Sheffield had experimented with chrome steel as early as 1871 when it was made for the first time at the Atlas Steel works.

The invention of alloy steels for tools is associated with the name of Robert Mushet, who in 1871 found that a manganese tungsten steel could be made sufficiently hard by allowing it to cool in the air. Previously all cutting tools had been made of carbon steel which had to be rehardened at frequent intervals by heating and quenching in water, at the risk of deformation or cracking. These new alloy tool steels had a tremendous influence on the engineering industry, for the new tools outlasted five or six of the old and allowed machine speeds to be doubled. Though Mushet's steel has now been superseded by high-speed tool steel all subsequent developments have been based on his work.

The next advance in alloy steels was in 1883, when Robert Hadfield (1859–1940), invented manganese steel. Other workers had discovered that manganese in steel made it harder, but also more brittle. Hadfield found that the addition of large quantities of manganese—10% or more—produced steel with remarkable qualities of toughness and ductility. Its toughness was increased when it was heated to 1050° C. and quenched in water—a treatment which makes carbon steel extremely brittle. It was also found to be non-magnetic. Manganese steel has another useful characteristic, for under impact it becomes harder on the surface while remaining ductile internally and hence is extremely valuable for all machinery and plant subject to abrasive action, such as railway crossings, excavators, dredgers, and the like.

Hadfield also discovered silicon steel, originally as a tool steel; but on investigating the magnetic properties he found that steels containing up to 5% silicon possess high permeability, high electrical resistance and low hysteresis loss, making them particularly suitable for the magnetic material in motor and generating armatures, transformer cores and other electrical equipment. Since 1907 silicon steel has been one of the basic materials of the electrical engineering industry and has made possible the development of more efficient and more powerful electrical equipment.

F. Hall of William Jessop and Sons, Sheffield, carried out research in 1889 into the effects of nickel on steel simultaneously with James Riley of the Steel Company of Scotland. The resulting nickel steel alloys rapidly established their value, particularly for armour plate.

In the early 1900s the original self-hardening tool steels produced by Mushet were superseded by what are known as 'high speed' steels. Developed by F. W. Taylor and M. White of the Bethlehem Steel Corporation of America, these steels exhibited cutting speeds of up to 500 ft. per minute, against 30 ft. per minute for carbon steel tools. They contain large amounts of the carbide-forming elements which serve to furnish wear-resisting carbides and resistance to softening at elevated temperatures. The typical composition is based on 18% tungsten, 4% chromium, and 1% vanadium, with 0·5% carbon and sometimes cobalt.

Stainless Steel. In 1913 H. Brearley (1871–1948), head of the Brown Firth Research laboratory in Sheffield, trying to find a remedy for the corrosion of rifle barrels using cordite, experimented with chromium steels and discovered stainless acid-resisting steel containing 13% chromium and 0·3% carbon. It was found later by B. Strauss and E. Maurer in Germany that the resistance to corrosion and the mechanical properties were improved by addition of up to 8% Ni and increasing the chromium to 18%. Steels based on this composition are widely used today, being popularly known as the 18–8 steels. The chromium content which confers resistance to corrosion also gives these steels resistance to oxidation and scaling at high temperatures, so that they are used as materials in the production of atomic energy, guided missiles, and gas turbine blades.

BIBLIOGRAPHY

1. CARR, J. C., and TAPLIN, W., *History of the British Steel Industry*, Basil Blackwell, Oxford, 1962.

2. *History of Iron and Steelmaking in the United States: The Metallurgical Society.* American Institute of Mining and Metallurgical Engineers, New York, 1961.
3. *Making, Shaping and Treating of Steel,* United States Steel Corporation, Pittsburgh, 1957.
4. *History of Technology,* Vol. IV. Iron and Steel, p. 98–117. Vol. V. Steel Industry, pp. 53–71. Oxford University Press, 1958.
5. DEARDEN, J., *Iron and Steel Today.* Oxford University Press, 1956.
6. DAVIES, MAX, *Story of Steel.* Burke Publishing Co. Ltd., London, 1950.

CHAPTER FIVE

THE MAJOR NON-FERROUS METALS

A HUNDRED years ago the range of non-ferrous metals extracted from the earth was in the main confined to copper, lead, tin and zinc, which had been produced from time immemorial by smelting the ores with charcoal or coal. Nickel and the light metals aluminium and magnesium are products of more recent times, the two latter having to await the advent of electrical generation before their production achieved any real prominence.

Due to advances in engineering and the science of metallurgy, tremendous progress both in the method and scale of smelting has characterized the last hundred years. The extent of this advance as shown by the growth in world production is given below, the figures for aluminium being particularly striking.

WORLD PRODUCTION OF MAJOR NON-FERROUS METALS* (TONS)

	1850	1875	1900	1950	1960
Copper	5,500	130,000	525,000	2,523,000	4,590,000
Lead	130,000	320,000	850,000	1,674,000	2,560,000
Zinc	16,500	165,000	480,000	1,819,000	3,510,000
Tin	18,000	36,000	85,000	162,000	179,700
Nickel	—	500	8,000	183,000	358,000
Aluminium	—	—	7,300	1,566,000	5,010,000
	170,000	651,500	1,955,300	7,927,000	16,207,700

In the 1850s Great Britain was pre-eminent in the extraction of copper, lead, zinc and tin, the Preface to J. Percy's famous treatise on *Metallurgy* (1861) stating that 'In no country are the operations of metallurgy conducted on so vast a scale as in Great Britain'. It is, however, sad to relate that this country's

* From *History of Technology*, Vol. 5, p. 73, excluding 1950 and 1960.

contribution to the advance which subsequently took place was negligible. Domestic mines were being worked out and the copper, lead and zinc smelters were no longer able to purchase supplies of rich ores from overseas mines, for smelters were being erected abroad on the mine site itself, especially in America. Electrochemical methods also were developed abroad where hydro-electric power was available at cheap rates, the costs of production and refining being much lower than in this country. This was particularly so in the case of copper and zinc. In face of this most of the British smelters went out of commission. The old smelters had done a good job, and their methods of smelting and purification have left their mark in metallurgical history.

COPPER

For more than 150 years during the 18th and 19th centuries, Swansea was the world's most important copper-smelting centre. Extensive collieries existed in the immediate neighbourhood from which an abundant supply of cheap coal could be obtained, and many of the smelters themselves owned and engaged in the working of the collieries. Swansea was also a seaport easily accessible to vessels conveying ore or products containing copper from many parts of the world, e.g. South America, North America, South Africa and Cuba. A large foreign trade was built up with ships carrying coal out of Swansea and returning with cargoes of ore from Chile, Peru, Australia, etc., more than 500 ships being engaged in the trade, their tonnage varying from 500 up to 1,000 tons. In its heyday (between 1850–60) some 600 furnaces were smelting copper, in addition most of the copper matte produced in the U.S.A. and elsewhere was being shipped to Wales for treatment.

A further factor was that the Swansea smelters had the privilege of pouring forth dense volumes of thick sulphurous arsenical smoke from chimneys into the atmosphere. Vegetation was destroyed with impunity by the gases, a right which would not readily be conceded in other parts of Britain. The inhabitants of Swansea generally seemed to be habituated to the inhalation of the smoke and to submit to the evil with unmurmuring resignation.*

By the end of the 19th century, however, copper smelting in

* J. Percy, *Metallurgy*, Vol. 1 (page 300), John Murray, London, 1861.

Swansea was virtually extinct, the chief causes of the decline being many and varied. The mines in Cornwall and Devon from which much of the copper ore was derived had become too deep and the lodes too poor to be economically worked. Costs of pumping water, mining ore and hoisting 200 or more fathoms could not be met from the declining grade of ore. The decline was also accelerated by the fall in the price of copper from £112 in 1860 to £35 in the 1890s, and by the discovery of vast deposits of low-grade ore in America, Chile and other parts of the world which could not sustain freight charges and necessitated smelting on the spot.

Welsh Process of Copper Smelting. Smelting up to the end of the 19th century was conducted in reverberatory furnaces and

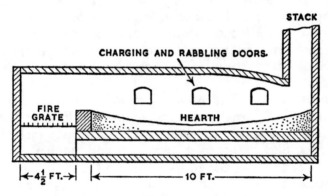

FIG. 10
WELSH REVERBERATORY FOR SMELTING COPPER ORES

consisted of a series of roasting and melting operations. The method illustrates the principles on which the extraction of copper is founded and is still based.

The reverberatory furnace (Fig. 10) in which the smelting was performed originated in Wales about 1800 and consisted of a hearth about 10 ft.–13 ft., built with fire or silica brick roof, clay brick side walls and a fused quartz bottom prepared by strongly heating sand over which some metal slag had been spread, the sand thereby being consolidated. A second layer of sand was spread over the cool surface and the process of heating with slag repeated until a total thickness of some 20 in. had been attained. The upper surface of the sand was shaped into a shallow cavity which gradually inclined from all sides to a taphole located on one side of the hearth. The furnace was charged with ore either through holes in the roof or through side doors. A fireplace about $4\frac{1}{2}$ ft. $\times$ $4\frac{1}{2}$ ft. was situated at one end of the furnace, being separated from the hearth by a fire-bridge. At the opposite end was a flue communicating with a high stack. Ores from different localities which varied from a low of 4% Cu to as much as 65% (cement precipitate from the Rio Tinto mines in Spain) were mixed to a uniform assay of 8–10% copper and submitted to the following operations.

1. *Roasting*. This was carried out in a separate reverberatory. The ore was stirred at intervals by rabbles through side doors, thereby preventing the ore from fusing and interfering with the efficiency of the roasting operation. Complete elimination of sulphur was not the aim, enough being left in to combine with the copper and iron in order to form a matte in a subsequent operation. When roasting was complete, which took 12–24 hours according to the nature of the material, the ore was raked out, cooled by wetting with water and wheeled to the ore furnace.

2. *First Fusion*. The roasted ore together with slag from operation No. 4 was introduced to the ore furnace through holes in the roof, the fire made up and the charge melted over 5–6 hours. Slag was then skimmed off and another charge of roasted ore and slag was introduced and the process repeated. The product 'coarse' metal was tapped out and granulated by causing it to flow into water contained in a pit in close proximity to the furnace. Preliminary roasting had burnt off part of the sulphur contained in the ore rendering (oxidizing) the iron into a form suitable for combination with silica in the gangue. During fusion in the ore furnace the whole of the copper combined with sulphur to form cuprous sulphide (Cu_2S), the remaining sulphur combining with iron to form FeS, the two uniting to form 'coarse' metal or matte. Hence the result of the fusion was that the copper had been concentrated into a product assaying 25–35% Cu, the gangue in the ore being eliminated as a ferrous silicate (FeO SiO_2).

3. The granulated matte was roasted with free access of air to oxidize the copper and iron in order that elimination of iron might be the more readily effected in the next operation.

4. *Second fusion or 'running for metal'*. The oxidized material

was melted with slags rich in copper oxide from the two remaining operations and any oxide copper ore available. The resulting products consisted of a slag composed of silicate of iron containing from 2–5% of copper (which formed part of the charge in operation No. 2) and a white metal composed of cuprous sulphide which contained about 85% of copper and which may be regarded as coarse metal (No. 2) devoid of iron sulphide.

The relevant reactions were

$$Cu_2O + FeS = Cu_2S + FeO$$

and

$$2Cu_2O + 2SiO_2 + 2FeS = 2FeSiO_2 + 2Cu_2S$$

the silica required to combine with the iron arising from the slags which were introduced with the charge. The white metal was tapped off into sand moulds.

5. *Roasting to blister copper.* The pigs of white metal were piled into the furnace and the fire regulated so that the pigs melted and trickled down the hearth of the furnace, being freely exposed to the air which circulated through openings in the side and roof of the furnace. When completely molten the slag was skimmed off to expose the surface of the metal to further oxidation. The charge was then allowed to cool somewhat in order to allow the molten metal to 'rise' or 'boil' due to evolution of sulphur dioxide.

$$2Cu_2S + 3O_2 = 2Cu_2O + 2SO_2$$

$$Cu_2S + 2Cu_2O = 6Cu + SO_2$$

The skimmings and boils were repeated several times until the metal was quiescent, when any slag was skimmed off and the metal run into sand moulds. All the slags which were rich in copper were collected and used in operation No. 4. Blister copper takes its name from the small craters produced on its surface by the escape of SO_2 during solidification.

6. *Refining.* This is the final operation, the object being to purify the metal by removing as far as possible such impurities as sulphur, arsenic, antimony, iron, etc. The removal of these impurities was essentially accomplished by their oxidation and absorption in slags.

From 6–10 tons of blister copper was charged to the furnace, melted, and kept exposed for about 15 hours to the oxidizing action of the air admitted to the furnace through the roof and side doors. Such oxidation was only partially achieved during the course of melting, so that when the charge was completely molten and the first slags removed, the main process of oxidation commenced. This was referred to as the 'flapping' stage and consisted in rabbling the bath with iron rabbles in order to expose the maximum amount of metal surface to the furnace atmosphere. The oxygen content at this period would be in the region of $0 \cdot 6$–$0 \cdot 9\%$ and was checked by taking a small sample and observing the rise or fall* of the surface during solidification. If it rose, the metal still contained sulphur and the oxidation process was continued. If it 'set' with a depression on its surface, it was satisfactory and ready for 'poling', the object now being to reduce the oxygen content of the copper to tolerable limits. This was accomplished by covering the bath of metal with anthracite or charcoal; and thrusting long green birchwood poles below the surface. The timber in contact with the copper was rapidly decomposed with vigorous evolution of reducing gases which combined with the oxygen contained in the metal.

Samples were taken from time to time, and when the 'set' was flat and the fracture salmon-coloured and of a fine silky appearance, the copper was ready for casting. Level set surfaces correspond with an oxygen content of $0 \cdot 025$–$0 \cdot 05\%$, this being the normal content of 'tough pitch' copper. The final stage in the process is that of casting and was hand-performed by means of long-handled ladles holding about 30 lb. weight of metal. It called for considerable skill, the 'knowhow' being no less important than it is today if unsound metal is to be avoided. When the copper was intended for rolling, from 3–12 lb. of lead per ton of copper were added just before ladling; the lead combined with any sulphur which might otherwise have caused brittleness and hence rendered the metal malleable. (Nowadays lead is regarded as an objectionable element in copper even in quite small amounts as it adversely affects properties and occasions cracking in hot rolling.) The whole process of smelting took from 70–100 hours. From 13–18 tons of coal costing 5/-

* The condition or set is essentially a phenomenon of the combination of oxygen with hydrogen to form steam. It is controlled almost entirely by the final oxygen content; it is not influenced by variation of the hydrogen content but is influenced to some extent by sulphur.

per ton were required to make one ton of copper, about half the quantity being consumed in the first and second operations of roasting and melting.

Although the operations given above number six, in reality the steps of the process were usually far more numerous, for a number of intermediate products and rich slags were made which required resmelting. However, the highlights in the above series of operations can be summarized as follows:

1. Preliminary roasting for partial elimination of sulphur.
2. Smelting in a reverberatory to concentrate the copper into matte with elimination of gangue as slag.
3. Production of white metal by oxidation of the FeS in the matte and its elimination as slag.
4. Production of blister copper by further oxidation.
5. Refining by removal of sulphur, arsenic, antimony, etc.

In essence these operations form the basis for the modern smelting of sulphide ore. The first step, however, is now omitted in all newly erected plants, the wet flotation concentrate being charged direct to the reverberatories. The matte which is the product of the reverberatory is poured into a vessel known as a converter and oxidized by blowing air through it. The resulting blister copper, after a preliminary fire refining is cast into anodes and refined by the electrolytic process.

In order to get an idea of the events which led up to modern smelting practice, it is instructive to consider the development of the reverberatory copper smelting furnace, which took place after the decline of copper smelting in Swansea.

To meet the growing demand for copper, stimulated especially by the expanding electrical industry, the copper reverberatory increased from the small 13 ft. Welsh furnace to the present 100–120 ft. continuously operated reverberatory capable of smelting 2,000 tons of ore per day (Fig. 11). This was essentially an American development, the first reverberatory to be built being that at Butte, Montana, erected in 1879 at the plant of the Colorado Smelting & Mining Company. This furnace had a hearth 14 ft. long by 9 ft. wide, using wood as fuel, and smelted about 10 tons of ore per day producing a matte assaying 60%

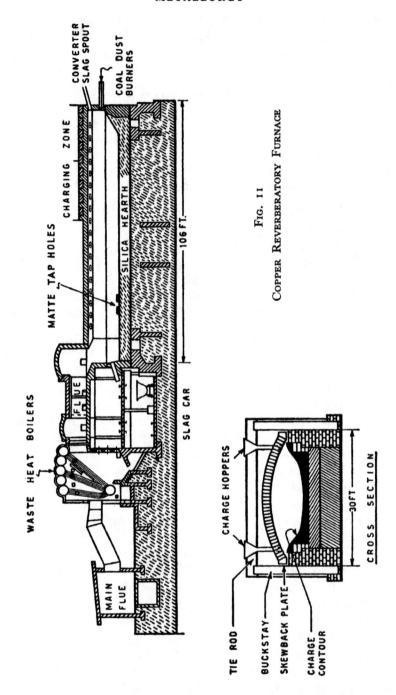

FIG. II

COPPER REVERBERATORY FURNACE

copper which was shipped to Swansea. Attempts to convert the matte to blister copper at Butte were given up on account of excessive cost, it being far cheaper to ship high grade matte to Swansea where labour skilled in the conversion was available at low wages. Furnaces remained small for the next ten years when they were about 14 ft. × 35 ft. and the fireboxes 5 ft. × 8 ft. These furnaces smelted 50–70 tons of ore per day with a fuel ratio of 3 tons of ore to 1 ton of coal. About 1890, furnaces were being built with longer hearths and relatively wide fireboxes which led to an intensification of heat in the region of the firebox.

About this period the practice was started of leaving matte in the furnace while the hot roasted calcines were charged more frequently. The maintenance of a large pool of matte in the furnace at all times with frequent charging of hot calcine was a milestone in the history of reverberatory practice, for it served to make the operation continuous. The decade 1890–1900 saw a growth in hearth dimensions to 19 ft. × 50 ft. and to fireboxes of $5\frac{1}{2}$ ft. × 10 ft. smelting of 120 tons of ore per day with a fuel ratio of 3 tons of ore to 1 ton of coal. Between 1900 and 1906 the Anaconda Copper Mining Co. built furnaces with hearths 19 ft. wide and lengths of 60, 85, 102, 112 and 116 ft. with results tabulated below:

SMELTING DATA FOR FURNACES OF DIFFERENT LENGTHS

Hearth Area (ft.)	Tons smelted per 24 hours	Fuel Ratio (Ore/coal)
19 × 50	121·7	2·75
19 × 60	190·7	3·94
19 × 85	234·1	4·13
19 × 102	266·0	4·31
19 × 116	270·1	4·19

ASSAY OF MATERIAL ENTERING AND LEAVING FURNACE

	Charge	Slag	Matte
Copper (per cent)	9·31	0·39	41·7
Silver (grms. per ton)	6·43	0·19	28·4
SiO_2	28·6	39·7	0·3
Fe + Mn	26·5	32·8	26·5
S	7·3	0·9	25·8

(*Trans. A.I.M.E.*, Vol. 106, 1933, p. 60.)

It will be noted that beyond a length of 102 ft. there was only a small increase in tonnage and no increase in fuel ratio. It would thus seem that a furnace of that length was about the economical limit. In 1902 utilization of the heat in the waste gases which left the furnace at a temperature of 1300° C. was effected at the Anaconda Works by the provision of waste-heat boilers. The boilers were of the Stirling type with vertical water tubes, the gases leaving the furnaces via an uptake flue and passing between the tubes, steam being generated at 220 lb. pressure and used for power generation, blowing engines and general purposes. These furnaces were still being fired by coal in fireboxes, but within a few years many smelters had substituted fuel oil and as a result of the more intensified combustion were smelting 500–600 tons of ore per day at a fuel ratio of about 25 : 29 U.S. gals. per ton of charge. This corresponds to 3·6 million B.Th.U. per ton of charge as compared to 5·6 million B.Th.U. when using coal. In 1913, International Nickel of Canada at the Copper Cliff Smelter succeeded in firing with pulverized coal and in 1914 Anaconda followed suit and were able to smelt 460 tons of ore per day in 20 ft. × 124 ft. furnaces with a fuel ratio of 6·2 tons of charge per ton of coal.

An important innovation was introduced in 1920 at Cananea in Brazil. The copper concentrates were so fine and gave so much trouble with dust losses during roasting that it was decided to charge the cold wet concentrates direct to the reverberatories, thus initiating the beginning of what is known as 'wet charging'. The advantage of this type of charging which today is practically universal is that the elimination of the dust in the reverberatory simplifies plant control, working conditions are improved, losses are reduced and the wear and tear on the furnace brickwork is lessened. Naturally a higher gross fuel consumption takes place than with a roasted charge, but this is offset by an increased waste-heat recovery together with fuel saving by dispensing with the roasting operation.

The development in size of the reverberatory necessitated by the increased demand for copper led to exacting demands on refractories. The increase in hearth area involved heavy stresses within the refractory lining, accentuated by the throughput of large tonnages and high temperatures necessary for economic operation. In addition, the lining is subjected to abrasion, mechanical erosion, corrosion through the chemical action of slags, fluxes and gases such as chlorine, sulphur dioxide and superheated steam.

Silica brick has been and still is widely used in all parts of the furnace, and in this respect has, indeed, rendered pioneer service in the expansion which has taken place in smelting. It possesses good load-bearing capacity at high temperature, constancy of volume and a high temperature of incipient fusion. Its resistance, however, to changes of temperature below a dull red heat is not very great and resistance to slagging is not all it might be owing largely to its porosity and acid character. In the early rever-beratories the hearth was constructed over an open space through which air circulated with the idea of keeping this portion of the furnace cool. The area was arched over and supported the hearth which was in the shape of a shallow pool enclosed by walls of firebrick supporting the roof. Later it was recognized that this space led to severe heat losses, and this error was accordingly rectified by giving the furnace a solid bottom of concrete or slag.

Nowhere are proper design and construction more important than in the roof. Among the factors which determine its stability are the materials of which it is built, the rise, thickness, length of span, furnace temperature and stability of buckstays and skewbacks. Movement of the latter was one of the most common causes of arch failure resulting from defective founda-tions or overheating of the metal plating. Another factor leading to failure of the roof is chemical action. Dusting created by charging the furnace, fume and gas from operation, and especially coal ash from the burners, are all extremely corrosive and react readily with the hot end of the brick.

A roof where silica brick is employed is always of the sprung arch type, a single layer of brick up to 20 in. in thickness being used. The span rests against [-shaped plates (skewbacks) running along both sides of the furnace and bolted to the buck-stays. These take up the thrust of the arch. As the roof is self-supporting over the whole span, its width is necessarily limited by the weight and crushing strength of the brick at the tem-perature of operation. This width for silica brick is about 25 ft. The roof is generally horizontal throughout its full length. In the early days the roof sloped steeply towards the front of the furnace in order to bring the hot gases into close contact with the bath. Modern methods, however, necessitate an adequate draught area in the front for the escape of products of combustion.

The latest development in roof construction is a mechanically suspended type of arch employing magnesite brick. Magnesite being more than 50% heavier is denser than silica and possesses a lower crushing strength and it has not yet been possible to maintain a sprung arch of magnesite brick under working conditions. The bricks are therefore suspended from horizontal supporting rods above the furnace.

Development of Copper Converting. Up to 1880 practically all copper matte produced in America and elsewhere was shipped to Wales for conversion to blister copper. Attempts had been made in America to reduce the matte in Bessemer type converters, but these failed, chiefly because the tuyère holes, which were in the bottom of the vessel, became clogged with chilled copper as soon as the reduction commenced, thus preventing air ingress. Pierre Manhes of France remedied this by changing the position of the tuyères and placing them some distance above the bottom so that there was space for the copper to settle underneath the blast. The first attempts to operate this type of converter likewise failed, for solidified matter still persisted in freezing in the tuyères. This difficulty was finally overcome by making openings in the air chamber opposite each tuyère so that a steel bar could be driven through to keep them open while the converter was in operation. These holes when not in use were kept closed by wooden plugs.

The first copper converter was installed at the Parrot works at Butte in 1884 and the practice of converting matte instead of shipping it to Swansea was started. This converter was a small egg-shaped vessel about 5 ft. in diameter lined with a mixture of crushed silica and fireclay and capable of being tilted for purposes of charging and discharging by a windlass and pulleys. Capacity was about 3 tons of matte. This first converter was eventually followed by a 10 ft. high, 6 ft. diameter converter, tilted hydraulically and served by a crane and holding from 7–9 tons of matte.

The introduction of the converter process was a very important innovation, for the conversion of matte to blister copper was done in a highly efficient manner, inexpensively, and could be carried out anywhere with relatively unskilled labour. The plant was compact and had a large capacity, 12 converters 10 ft. high, 6 ft. diameter, producing about 5,000–6,000 tons of copper per month.

The operation of converting is carried out by transfer of matte from the reverberatory to the converter by means of a ladle and an overhead crane. The charge is introduced through the mouth of the converter, the blast turned on and the converter turned up. The conversion reaction takes place in two stages corresponding roughly to stages 4 and 5 of the old Welsh process (page 131). During the first stage the iron sulphide in the matte is oxidized to FeO which combines with the siliceous lining of the converter to form slag, the heat necessary for the reaction being provided by the oxidation:

$$2FeS + 3O_2 = 2FeO + 2SO_2 + 240 \text{ K cals.}$$

The progress of the operation may in general be gauged by the colour of the flame issuing from the mouth of the converter. During the first part of the blow the flame is reddish, but soon changes to green which continues for about 45 minutes. As the oxidation of the iron nears completion the colour of the flame turns to pale blue. At this stage the converter is turned down, the blast turned off and the iron oxide slag poured off. More matte is then added, the converter turned up and the operation repeated. When sufficient white metal (Cu_2S) has been accumulated, the second stage commences, the sulphur in the white metal being oxidized, the oxidized copper then reacting with the remaining white metal to produce metallic copper and sulphur dioxide (page 131). During this period the colour of the flame changes from pale blue to reddish brown, and constant punching of the tuyères is usually required to keep them free from solidified copper. When the blow is considered to be finished, the converter is turned down, the blast shut off and the copper poured into a ladle for casting. The slag from the converting process containing up to 5% copper is transferred to the reverberatory furnace for reclamation of its metal content. The cost of the lining was a considerable item of expense for it was rapidly corroded by its combination with the FeO contained in the matte, for this lining was the sole source of silica supply needed to flux the FeO. The life of an acid lining was limited to 5–7 charges and the expense of constantly renewing was very heavy. Many attempts were made to find a more durable lining. Finally, in 1910, W. H. Peirce and E. A. Smith, after many years of research at the Garfield Smelter, Utah, succeeded in accomplishing the task by the use of a lining composed of refractory bricks of basic magnesite which was not attacked by the iron. In addition they radically changed the shape of the converter from an upright squat vessel to a cylindrical form lying in a horizontal position encircled by flange rings resting on rollers, capable of being rotated on its axis by electrical power for purposes of charging and emptying. The silica needed for fluxing the FeO was added with the charge.

The present day converter is commonly 13 ft. in diameter and 30 ft. long, and holds about 80–100 tons of matte at a charge and can produce 100 tons of blister per 24 hours from a 45% matte.

The advantages of this type of converter are that vessels can be made much larger, lower grade mattes can be treated, and siliceous ores containing copper and precious metals can be employed as fluxing material yielding up their metal content at no extra cost.

As normally smelted, blister copper rarely exceeds 99·3% Cu and is frequently as low as 98·6%, the most common impurities being iron, sulphur, lead, arsenic, antimony, nickel and the precious metals gold and silver. As these residual impurities adversely affect the electrical conductivity and mechanical properties, a refining treatment is almost invariably required.

Two methods are available, fire refining and electrolysis.

To be amenable to fire refining, blister copper should contain no precious metals and the impurities should be low enough to yield good quality commercial copper. As the amounts of copper conforming to these conditions is at the present time about 12–15% of the world's total of primary copper, it is apparent that at least 85% of the world's copper is currently obtained by alternative processes, of which electro-deposition contributes by far the largest total. Electrolytic refining not only enables almost complete recovery to be effected of the precious metals, but also permits separate recovery of many of the impurities such as selenium, tellurium, nickel and bismuth. In addition an excellent quality of copper is produced.

ELECTROLYTIC REFINING

The introduction of electrolytic refining as an industrial operation was one of the most significant advances in copper technology, for not only did electrical generators make the process possible but the pure copper thus produced made possible the enormous expansion in electrical equipment.

Hitherto the chief outlet for copper was for the manufacture of domestic pots and pans, sheathing of ships' bottoms and fireboxes for locomotives. With the advent of pure copper, with its high electrical conductivity and the readiness with which it could be drawn into rod and wire, the chief demand arose from the electrical industry.

The electrical conductivity of copper varies with the presence of various impurities, phosphorus, silicon and arsenic for instance being highly detrimental; hence the electrical industry needed a high purity copper which was capable of production by electrolysis but not by the fire-refining methods then in use. One of the first patents relating to the electrolytic refining of copper was taken out in 1865 by J. B. Elkington a Birmingham electro-plater. Blister copper was cast into anodes 18 in. square $\times$ $\frac{3}{4}$ in. thick with cast on lugs for suspension in tanks from conductor rods. Spaced in between were thin cathode copper plates $\frac{1}{32}$ in. thick on which pure copper was deposited from the anodes. A saturated solution of copper sulphate formed the electrolyte. Special note was made in the patent of the possibility of the recovery of precious metals from the anode slimes.

The first electrolytic refinery was constructed in 1869 at Pembrey, near Swansea, in the copper smelting plant of Mason & Elkington. Many other electrolytic plants were soon operating in the Swansea area. The economic operation of these plants was based on the high value of the silver and gold recovered from the anode slimes when using blister coppers derived from ores rich in precious metals, mined chiefly in Spain and South America. Electrolytic refining was continued in Wales until about 1912.

The first electrolytic refinery on the American continent was at Laurel Hill, New York, in 1892. This refinery was also designed to operate with blister copper, which was purchased from various sources, a high precious-metal content being the basis of choice. Later, with the increasing demand for copper of high conductivity, electrolytic refining plants were installed by most major copper producers, and the refinery tank-house was eventually to become a principal world source of precious metals, as well as of some of the other less common metals, including bismuth, cobalt, selenium and tellurium.

The practice of refining blister copper direct ceased by the end of the century, because iron and other impurities contained in it produced a rapid deterioration of the electrolyte. Instead, the metal was first subjected to a preliminary fire-refining and the tough pitch copper cast into anodes. After electrolytic refining, the cathodes were for the most part remelted and subjected once more to the cycle of oxidation and poling with green wood to tough pitch. Other modifications in electrolytic refining have been of a minor character, designed in the main to secure closer spacing of the electrodes and so to increase current efficiency, upon which the economic success of the operation so largely depends.

In the modern electrolyte copper refinery the anodes, measuring approximately 36 in. × 36 in. × 1½ in. and weighing 400–600 lb., consist of blister copper which has been partially refined in the anode furnace. The anodes are fed to the electrolyte tanks, and one full tank is charged at a time, 30–40 anodes being positioned by an overhead crane. They are suspended in the tanks by means of two lugs and are spaced 3–4 in. from centre to centre. Between the anodes, starting sheets are hung by hand by two loops of the same material from copper rods. These are thin copper sheets of a thickness about 0·03 in. and weigh 9–12 lb. and are prepared in cells known as stripper or starter cells especially set apart for the purpose.

The electrolyte is primarily a solution of copper sulphate with a copper content of 3–4% with 12–15% free sulphuric acid. By electrolytic action the anode copper is dissolved and deposited in a pure condition on the cathode starting sheets, the soluble impurities such as nickel, iron, arsenic and antimony being dissolved in the solution, the insoluble impurities such as lead, selenium, tellurium and the precious metals gold, silver and platinum remaining on the surface of the anode, or else sloughing off to the bottom of the tank as slime. In order to promote the speed and efficiency of deposition, the electrolyte is first heated in special tanks provided with lead coils through which low pressure steam is passed. Electrical resistance of the solution constitutes about 50% of the total resistance of the circuit and since the conductivity of the solution is a function of temperature as well as of free acid (and various soluble impurities) an economic balance has to be struck between these various factors. The practice is to heat the solution to 50–60° C. and circulate it through the tanks at from 3 to 4 gals. per minute.

Additives such as glue and bindarene are added to the electrolyte to promote smoothness and density in the deposited copper. In some refineries there is a tendency for the copper content of the solution to build up beyond the desired maximum. As this is not conducive to efficiency, the solution is passed through tanks known as liberators, which use hard lead anodes and ordinary cathode starting sheets, the plating out of copper depleting the solution of the metal.

The tanks and internal electrical circuits are generally arranged according to the Walker multiple system, with the tanks in series, the electrodes (i.e. the anodes and cathodes) in parallel. A current of 10,000–12,000 amps is passed through the tanks, the circuit voltage required varying from 75–150 volts depending on the load. Current density is maintained at 12–15 amp/sq. ft. in the depositing tanks and a little higher in the starting sheet tanks. Current efficiency averages 90%.

Cathodic deposition is usually complete after 14–15 days and as the anode life is 28–30 days, two crops of copper per anode are obtained. Overhead cranes lift all the cathodes from one tank in one load, when they are washed with hot water to remove solution and then delivered to the storage building ready for transport to the wire-bar furnace or cathode shears.

In order to maintain the efficiency of the system a certain amount of electrolyte is removed from the tank house each day. The principal solution impurity is nickel which markedly increases the resistance of the solution. The solution is first decopperized by electrolysis in tanks using lead anodes, the nickel bearing solution then being evaporated under vacuum when nickel sulphate crystallizes out. Clear acid is separated from the crystals, filtered and returned to the tank house. The nickel residues are shovelled into centrifuges equipped with rubber buckets where they are washed free of acid and dried.

Electrolytic slimes are periodically flushed from the tanks and pumped to the refinery for recovery of precious metals, selenium, tellurium, etc.

ALUMINIUM

Too young for any legendary associations, the metallurgy of aluminium began with its discovery by Sir Humphry Davy in 1807. It took, however, over 50 years for aluminium to advance from experimental production in the laboratory to the first bar of metal, so amazingly light that it seized the imagination of Emperor Napoleon III, who ordered the new metal to be used instead of silver for his table service. With an eye on its potential strategic advantages in weapons of war, he encouraged research into large-scale methods of manufacture.

Chemical methods of extracting the metal from bauxite (the oxide ore of aluminium) were at first used to isolate the element. Later on, electrolytic methods of production were investigated, but with primary cells as the only source of electric power, this approach was not very fruitful until—in 1870—the dynamo appeared. It paved the way for more research and in 1886 aluminium emerged as a commercial metal based on a method of electrolysis invented simultaneously in France and the United States.

ELECTROLYTIC PRODUCTION

The first attempt to produce aluminium electrolytically was made in 1807 by Sir Humphry Davy. Nearly fifty years later R. W. Bunsen and St. Claire Deville by passing an electric current (obtained from a battery) through fused aluminium sodium chloride contained in a porcelain crucible obtained buttons of aluminium.

Apart from the fact that the only source of current at that time was from batteries, attempts to manufacture aluminium upon these lines were beset with other difficulties. No material could be found suitable for making crucibles which when heated were not acted upon by the fused haloid salts or by the metal itself at the temperature employed, impure aluminium resulting. Crucibles of silica, clay, graphite and metal were all attacked by the chloride salt.

Davy, in isolating sodium and potassium by electrolysing caustic soda and potash, had stated that 'he had attained his object by employing electricity as the common agent for fusion and decomposition'. Hence, it would appear that in order to make possible the electrolysis of aluminium, heat must be generated within the fused mass.

The first application of the electric current for the reduction of alumina was due to the Cowles brothers who in 1884 set up furnaces at Cleveland, Ohio, and also a few years later at Stoke-on-Trent for the production of aluminium alloys. The process was, however, not electrolytic but electro-thermal, for the current was used solely to generate the high temperature necessary to effect reduction.

The furnaces were rectangular in cross-section, 5 ft. long and $2\frac{1}{2}$ ft. wide, constructed of fireclay, with two carbon electrodes which entered at the side walls and were inclined at an angle to one another. Current was supplied from a dynamo of 400 h.p., the potential being 60 volts at 600 amps. Bauxite, charcoal, and the metal forming the alloy were placed in the furnace, the current turned on, resistance to the passage of the current offered by the charge generating heat which effected reduction of the alumina by the carbon, the liberated aluminium immediately combining with the molten metal present to form an alloy. This collected at the bottom of the furnace and was tapped out at the end of the operation which lasted for about one hour. Copper was commonly used, the alloys containing 15–40% aluminium.

Expenditure of energy necessary to produce 1 kilo of aluminium was found to be on the average 40 h.p., the theoretical energy required being only 8·9 h.p. Thus the electrical efficiency of the process was very low indeed.

For many years the aluminium alloys made by the Cowles process were sold at a lower price ($5) per lb. of contained aluminium than pure aluminium itself. The works at Stoke-on-Trent ceased production in 1893, the process then being obsolete. It will be noted that in the process a metal (with which the alloy is to be made) is included in the charge to combine with the liberated aluminium. If no metal is present, alumina on reduction with carbon yields the carbide ($2Al_2O_3 + 9C = Al_4C_3 + 6CO$) hence production of aluminium by this reaction is not feasible.

In 1886 C. M. Hall in America and P. L. Heroult working independently in France discovered simultaneously* that

* A further coincidence was that both men were born in the year 1863 and both died in 1914.

molten cryolite would dissolve alumina and that this solution when electrolysed would deposit metallic aluminium.

The solvent, cryolite, was not appreciably affected by electrolysis and hence could be used indefinitely; further, its low density permitted the aluminium to sink through it to the bottom of the cell and thereby be protected against oxidation. The method eventually brought the price of aluminium down to 1s. 6d. per lb. and paved the way for the emergence of aluminium as a commercial metal.

Not possessing sufficient money to establish himself on a commercial footing, Hall first approached the Cowles brothers and entered into an agreement to demonstrate his process at their works. After a year or so they informed him that they were no longer interested in his method. Hall next approached an acquaintance named Cole who realizing the possibility of the method formed a company known as the Pittsburgh Reduction Co. (now the Aluminium Company of America) and commenced operations in 1889 in a couple of rooms.

The electrolytic cells were of carbon-lined cast iron shells, measuring 24 in. long, 16 in. wide, 20 in. deep. Carbon anodes measuring 3 in. in diameter and 15 in. long (6–10 in number) were suspended in the cells from an overhead copper busbar. The cells themselves constituted the cathode, being connected to the negative terminal. Each cell held about 200–300 lb. of electrolyte composed of alumina dissolved in cryolite. Each two pots were arranged in series and took 1,700 amps at 16 volts, output averaging 50 lb. of aluminium per day.

The first commercial exploitation of Heroult's process took place in 1887, not in France, but at Neulaissen in Switzerland by the Société Metallurgique Suisse utilizing electrical power generated at the Rhine Falls. The process and equipment were essentially the same as those operated by the Pittsburgh Reduction Company.

PREPARATION OF ALUMINA

Bauxite, the only commercial source of the metal, takes its name from Les Baux, in Southern France, where it was originally mined. It is an impure hydrated form of alumina, possessing the following approximate composition:

146

Al_2O_3 54–60%; SiO_2 2–6%; Fe_2O_3 1–20%; TiO_2 0·5–3%; H_2O 14–25%.

The table below gives the world output of bauxite for the chief producing countries.

(Production in thousand/tons)

	1918	1960
British Guiana	4·2	2,471
Dutch Guinea	—	3,400
France	145·0	2,006
Hungary	—	1,170
U.S.A.	615·4	1,998
Jamaica	—	5,745
Others	198·2	10,270
	962·8	27,060

The extraction of aluminium of a suitable high purity from bauxite demands the preparation of alumina of a high order of purity, and hence the purification of the bauxite is the first operation.

The process in use today was invented in 1887 by C. J. Bayer in Germany and has survived mainly because of its relative simplicity and the purity of its product. Many other methods have been proposed or operated from time to time, of which the following are typical examples:

Deville–Pechiney Process. This method, introduced in the 1860s, was formerly the principal method used for production of alumina. The finely ground bauxite was mixed with soda ash in the proportion of 3 : 1 and then sintered in a reverberatory furnace for 2–3 hours at 1100° C. The soda ash combined with the alumina to form sodium aluminate, carbon dioxide being evolved.

$$Al_2O_3 + Na_2CO_3 = 2NaAlO_2 + CO_2.$$

The iron forms sodium ferrite and the silica, sodium aluminate silicate. On digesting with warm water the soluble sodium aluminate goes into solution, the ferrite precipitating as ferric hydroxide. The sodium aluminate silicate is insoluble and hence represents a serious loss of both soda and alumina, approximately 1 lb. of Al_2O_3 and 1 lb. of soda being lost for each pound of SiO_2 in the bauxite. (For this reason it is customary to penalize the mine for each unit of silica in excess of some scheduled amount.) The liquor was allowed to settle, the supernatent liquor being withdrawn and filtered, the clear liquor then being saturated with carbon dioxide gas at 70° C. when aluminium hydroxide precipitates.

$$2Na AlO_2 + CO_2 + 3H_2O = 2 Al(OH)_3 + Na_2CO_3$$

the CO_2 gas being derived from the reverberatory gases.

The filtered precipitate of aluminium hydroxide was dried and calcined in a rotary furnace at 1000° C. to yield a product containing 98% Al_2O_3. The solution of sodium carbonate obtained in the above reaction was concentrated by evaporation to recover the soda for re-use in the reverberatory operation.

BAYER PROCESS

This process, introduced by C. J. Bayer in 1887, in its essentials is still today the most widely used of all methods for the production of alumina. The process requires less expenditure on fuel, wages and repair than other processes and in addition yields a product of high purity.

In this process the bauxite is digested with a hot caustic solution under pressure, the alumina dissolving to form sodium aluminate. After separation from the insoluble residues, aluminium trihydrate is precipitated from solution and calcined to produce pure alumina.

The crushed ore is dried by calcining in a rotary kiln followed by fine grinding in ball mills. The ground ore is then agitated with 30% caustic soda and pumped continuously to digestors, steam at 125 lb. pressure providing the super-heat necessary for dissolving the alumina. After digestion the pulp passes to pressure release tanks where the pressure is reduced to atmospheric, the temperature dropping from 160° C. to 100° C. Recovery of the sodium aluminate liquor and the disposal of the insoluble impurities known technically as red mud is accomplished by a process of continuous thickening in thickeners, insulated and covered to conserve heat. Underflow from the last thickener containing the insoluble residue is filtered on a pressure filter and the cake discarded. Thickener overflow is clarified for removal of suspended matter and transferred to precipitation tanks, aluminium hydrate from a previous batch being added, which 'seeds' out new hydrate from the pregnant liquor. It is added

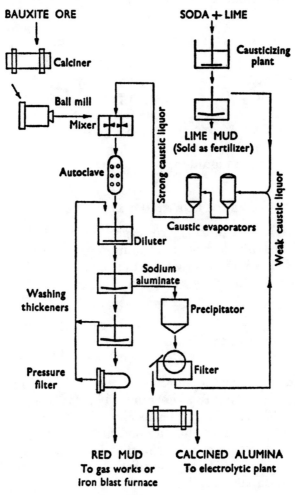

FIG. 12

ALUMINA PRODUCTION BY MODERNIZED VERSION OF BAYER PROCESS

in about the same quantity of new hydrate that the liquor is expected to yield. Stirring continues for 24–48 hours, when the precipitate is filtered off, washed and calcined at 1100° C. in rotary kilns and discharged to pipes surrounded by air delivered by fans where it is cooled down to atmospheric temperature.

The dilute caustic soda filtrate from the precipitate filters is concentrated in multi-effect evaporators and returned for re-use. Assays of the products are as follows:

ASSAY OF RED MUD AND ALUMINA

	Red Mud	Calcined Alumina
Al_2O_3 %	15–25	99·5
SiO_2 %	10–15	0·03
Fe_2O_3 %	40–60	0·02
TiO_2 %	2–6	—
Na_2O %	5–10	0·30
Loss on ignition	5–10	0·15

ELECTROLYTIC REDUCTION

The Hall–Heroult process for the production of aluminium is accomplished by electrolysis of alumina in a bath of molten cryolite. Quantities required for the production of one ton of metal are approximately as follows:

Alumina	2 tons
Cryolite	Very small consumption
Coke, pitch and tar for anodes	0·6 tons
Electricity	18,000 kWh.

Cryolite is a double fluoride of aluminium and sodium, Na_3AlF_6, and contains when pure 13% Al; 33·3% Na and 53·6% F. There is only one locality, Ivigtut in Greenland, where it occurs in any quantity and although huge quantities were shipped in the early days of the aluminium industry, it is now prepared synthetically by neutralization of hydrofluoric acid with soda and alumina.

$$3Na_2O + Al_2O_3 + 12HF = 2Na_3 AlF_6 + 6H_2O$$

In the solid state cryolite is denser than aluminium, the specific gravities of cryolite being 3·0 and aluminium 2·7, and in the molten state cryolite is 2·1 and aluminium 2·34. Molten cryolite when saturated with Al_2O_3 has a gravity of 2·15, and hence as molten aluminium is heavier it settles to the bottom of the electrolytic bath when liberated, which is very convenient for cell control, for otherwise it would float and upset cell operations. Cryolite melts at about 980° C., but its fusion point is lowered by the addition of Al_2O_3, thus when 10% is present it is 930° C. Calcium and aluminium fluoride are generally added to the bath in order to lower the fusion point, a typical bath composition being 86% cryolite, 6% aluminium fluoride, and 8% calcium fluoride.

Anodes are composed of carbon and take one of two forms, batch or continuous. In the former the anodes are composed of petroleum coke (the residue left after the distillation of petroleum) which has been heated to a high temperature in order to expel valuable matter and increase its density and conductivity. The material is then ground, mixed with a binder of pitch, and either extruded into cylindrical shape or compressed into cubes of 12–20 in. sides, weighing up to 200 lb. and baked. Embedded in the anodes are copper bars for contact with the cell busbar. Up to twelve of these anodes form the complement of one cell. Their life is very short, the quantity consumed in the process being nearly equal to that of the aluminium produced. They are suspended in the cells from racks motivated by gear wheels, each being capable of individual adjustment and replacement.

During World War One, C. W. Söderberg, a Norwegian engineer, developed an anode which possesses the merit of being continuously renewed as it is consumed. It was originally applied to the electric furnace production of ferro-alloys, but was soon adapted to aluminium production, for in addition to being continuous it afforded efficient cell arrangement and showed economies in construction, operation and materials. As originally conceived it composed a cylindrical casing extending into the furnace into which a carbonaceous mixture was continually fed. In its application to the aluminium cell the shape was changed to that of a hollow rectangular casing, about 4 ft. wide, 10 ft. long and 2–4 ft. in depth, composed of aluminium, supported structurally above the cell, and continuously filled with a hot mix composed of 70% coke and 30% pitch as binder. The electrode projects into the cell and as its lower edge is consumed, volatile matter is given off and the mass baked hard into a continuous electrode. It may be lowered or raised by a hand-wheel. Electrical contact with the anode is made by

151

a row of iron pins flexibly connected with the busbars above and inserted through the aluminium casing into the carbon mix before it is baked. As the anode is consumed the lower line of pins is withdrawn and replaced at a higher level. Usually the casing projects through an opening in the staging above the cell, new sections being added as required. One Söderberg electrode is used per cell.

It requires about 10 units of electricity to produce 1 lb. of aluminium. Ten units will keep the average two-bar electric fire burning for 5 hours, and will cost (in Great Britain) about 1s. 6d. One pound of aluminium today (1962) is sold for 1s. 8d., showing how important in its production is an abundant supply of cheap power, and why most of the aluminium metal consumed in Britain is imported.

<h2>THE CELL</h2>

The electrolyte cell consists of a rectangular steel shell insulated with firebrick and lined with a carbon mix which has been rammed into place. Embedded in the lining are iron plates or bars which project from the cell and are connected to the negative terminal. The lining thus acts as the cathode and also forms the container for the electrolyte and metal. A taphole is provided at the bottom of one of the long sides. The size of the cell depends on the amperage used, a cell taking 30,000 amps would measure 15–18 ft. long, 7–9 ft. wide, and 5 ft. high. The output of such a cell would be approximately 600–700 lb. of aluminium per day. Usually 100 cells or so are connected in series corresponding to an e.m.f. of 600 volts. With 10 such units the daily output would be approximately 300 tons per day.

For ease of working the cells are usually set below floor level with the top of the cell 15 in. or so above the ground.

The electrolyte consisting of molten cryolite together with added calcium and aluminium fluorides is maintained at 950° C., alumina being added to give an approximate 6% solution. Decomposition takes place according to the following reactions:

$$2Al_2O_3 = 4Al + 6O$$

the liberated cathodic aluminium sinking to the bottom of the cell, oxygen being given off at the anodic surface, where it combines with the carbon forming CO and CO_2. Cryolite is reasonably stable, and does not decompose to any great extent. In operation a cell contains a bottom layer of molten aluminium several inches in depth, an upper layer of fused electrolyte, a crust of solidified electrolyte and alumina forming over the top. Electrolysis proceeds until the concentration of alumina has been reduced to a low figure at which point the so-called 'anode effect' occurs, normal operation of the cell ceasing. At this point the electrolyte no longer wets the anode, a gas film—due to polarization effects—forming over the anode, which being electrically insulating causes a rise in voltage of as much as 50 volts, indicated by the lighting of an incandescent bulb connected across the terminals. The operator then stirs in a fresh charge of alumina which has previously been spread over the solidified crust so that it may be heated and dried if necessary. The addition of fresh alumina and vigorous stirring destroys the anode effect, the voltage dropping back to normal and electrolysis proceeding evenly for two hours or so when the anode effect reappears.

The anode is consumed in the process with evolution of carbon monoxide which agitates the bath maintaining the alumina long enough in suspension for it to dissolve in the electrolyte. Escaping through the crust the monoxide then burns in contact with the air. Molten aluminium is drawn off from the cell at convenient intervals, being conveyed to the casting department where it is cast into pigs. Alternatively, it may be tapped into a holding furnace which permits fluctuations in the composition of successive taps to be smoothed out.

Operating details are given below:

ELECTROLYTIC REDUCTION DATA

Current employed, amps	40,000–80,000
volts	4·5–6·5
Anode current density, A/sq. ft.	650–750
Current efficiency, %	85–90
kWh/lb. Al	10
Electrode consumption	
lb. per lb. of Al produced	0·5–0·6

REFINING

Aluminium obtained from electrolysis normally assays 99·5% Al, 0·2% Si, 0·15% Fe, and although suitable for alloying, many industries, notably the electrical and food industries, require metal assaying 99·9% and better. To achieve this high purity a refining process is required.

A. Hoopes, in an American patent issued in 1901, proposed an electrolytic three-layer process in which the impure aluminium alloyed with a heavy metal such as copper contained in a cell of appropriate design, was made to form a molten anode layer below an electrolyte, the refined aluminium rising through the electrolyte upon which it floated. Electrical contact was established with the anode layer through a carbon lining at the bottom of the cell and to the top cathode layer by graphite electrodes.

In the Hoopes process, the electrolyte consists of a molten mixture of cryolite and aluminium fluoride with barium fluoride added to give the necessary density, the approximate composition being:

Aluminium fluoride		30–38%
Sodium	,,	25–30%
Barium	,,	30–38%

The density of this electrolyte at the working temperature of 1000° C. is 2·5 g/cc. Pure aluminium at this temperature has a density of 2·30 and hence will float on the electrolyte. The 25% copper aluminium alloy employed has a density of about 2·8 g/cc. which is sufficiently high to ensure that it will not float but remain submerged beneath the electrolyte.

The cell consists of an open-topped steel shell with a lining of refractory bricks, the bottom being of carbon blocks through which electrical connection is made to the molten anodic layer of impure aluminium alloyed with copper. In order to prevent short circuiting between the anode and cathode through the hot lining the cell is divided into two electrically insulated sections arranged with water-cooling at the junction. By virtue of this cooling effect a thick crust of aluminium is built up on the side walls which serves electrically and thermally to insulate the molten electrolyte from the shell. The impure aluminium to be refined is fed at regular intervals to the anode copper alloy through a graphite tube, and from this layer pure aluminium is liberated and ascends to the top layer, from which it is periodically ladled off. The cell produces metal of about 99·9% Al and operates at 20,000 amps and 5–7 volts.

Compagnie A.F.C. Process. A process using a low temperature electrolyte was patented in 1934 by the Cie de Produits Chimiques at Electrometallurgiques Alais, Froges et Carmarque —the French aluminium producers—and has been used commercially since that time for the production of 99·99% metal. The electrolyte consists of:

Aluminium Fluoride	23%
Sodium Fluoride	17%
Barium Chloride	60%

The cell operates at about 740° C. Decreased reactivity of the electrolyte at the lower temperature enables water-cooling of the cell to be dispensed with. Thermal losses are reduced and operating conditions in general are rendered more amenable to control. The process still involves a high energy consumption of about 10–12 kWh/lb.

ALUMINIUM ALLOYS

Pure aluminium has a low strength (4–5 tons/sq. in. as cast) for engineering purposes; hence improvement of its properties by alloying so that wider use could be made of its lightness and durability was a first consideration. The earliest alloys were binary alloys, i.e. containing only one additional element, such as the aluminium-manganese alloy (which is still widely used for press work, spinnings and panelling) and the alloys of aluminium with copper, silicon or magnesium; later, more complex alloys were developed. The number of alloying additions to aluminium is restricted, however, to a few metals, the total alloying content of most wrought alloys being less than 6%.

Wrought Alloys. Aluminium really began to achieve industrial importance in the period 1900–14. One of the most important developments was the discovery of age-hardening by the German metallurgist, A. Wilm, in 1907. Wilm had been attempting to harden an aluminium alloy containing copper and magnesium by heating it to 500° C. and then quenching in water (following steel-hardening procedure). The experiment was unsuccessful—the alloy still remaining soft. Some days later having occasion to re-test the alloy Wilm found that it had become harder and on further investigation he discovered that the strength had also increased. Wilm then carried out a series of tests on the effect of storing the alloy for different periods (at room temperature), after heating and quenching. He found that in fact the strength improved on storing and reached a maximum after four days of about 25 tons/sq. in., the properties then becoming stable. This chance discovery opened up the way for the development of aluminium alloys as structural materials with a strength nearly equal to that of mild steel.

The discovery was the basis of the first Duralumin patent taken out in 1909, the word Duralumin being derived from the Aluminium Works at Duren in Germany to which Wilm gave the sole right to work his patents. At about the same time a research team at the Royal Arsenal, Woolwich, working independently on other alloys, made a similar discovery which however was not developed commercially.

In 1911 a beginning was made in Birmingham on the production of the first high-strength alloy based on Wilm's patents. The production of the Duralumin-type alloy coincided with the need for a strong light material for aircraft, and from that time the development of the high-strength alloys has been mostly linked with aircraft. Another important step was the development at the National Physical Laboratory, Teddington, during World War One of the 'Y' alloy. This was the first of a group of age-hardened alloys containing nickel, copper and magnesium, which retained their strength at high temperatures. The alloys have special applications for aircraft and motor engine components such as pistons, and are available in both wrought and cast forms.

The highest strength given by aluminium alloys is in the range 35–40 tons/sq. in.; these contain as their main alloying constituents, magnesium and zinc, with a smaller amount of copper.

Cast Alloys. Most of the earliest uses of aluminium were in the form of castings. Gilbert's Piccadilly Circus statue 'Eros', a familiar London landmark, was cast in aluminium in 1893, the supporting leg being of solid aluminium, the remainder being an assembly of hollow castings.

The strength of the pure metal when cast is too low for general engineering use, but as with the wrought material this is improved by the addition of other metals. Of the early cast alloys, those containing copper were most notable. In the early 1920s, A. Pacz found that extra strength and ductility as well as excellent running of the molten metal in the foundry were given by addition of silicon, particularly if 'modified' by addition of small amounts of sodium. This material known as Alpax has established itself as a standard material.

The casting alloys most widely used today contain both silicon and copper as alloying elements.

Output and Usage. From the end of the last century, in a period of little more than sixty years, the entire commercial development of aluminium has taken place, production of aluminium having expanded two hundredfold. Today world production of aluminium is greater than that of any other metal with the exception of iron and steel:

WORLD PRODUCTION OF ALUMINIUM
(thousand long/tons)

	1951	1961
United Kingdom	28	32
Europe	328	878
Asia	43	175
Africa	—	43
Australia	—	13
N. America, Canada	400	590
„ U.S.A.	747	1,703
Eastern Bloc	233	1,000
Total	1,779	4,434

The low weight of aluminium and its alloys in relation to its strength, its good resistance to corrosion and its ability to conduct electricity have been the main causes of the rapid growth of the aluminium industry. These properties offer important advantages in the construction of aircraft, road and rail vehicles, and in the construction of electrical conductors for carrying bulk power. Other important fields are the engineering and building industries and food trades, the following table

showing the present field of industrial usage in the United
Kingdom.

	Per cent
Transport, road and rail	22
Packaging	12
Building and construction	8
Electrical engineering	10
Domestic appliances	8
General engineering	6
Exports, semi-fabricated	10
Aircraft and miscellaneous	24

At the present time the aluminium industry is breaking much
new ground, e.g. plastic-covered aluminium sheet, high-
temperature alloys, and the next few years hold promise of
further and interesting developments in many spheres of
application.

ZINC

Historically, the oldest use of zinc was in brass, which was first
made by smelting mixed ores of zinc and copper, but zinc
smelting itself probably began in the 13th century in India
and then spread to China. The metal was brought to Europe
by Portuguese navigators in the 17th century and was first
made in Britain about 1720 by Dr. Lane in his Swansea copper
works. Production on a commercial scale began in 1740 when
William Champion (1709–89) set up a smelting works at
Bristol. The method was conducted by 'distillation per descensum', the zinc distilling and being conducted downwards to be
condensed and collected in trays placed to receive it. The reduction was effected in fireclay pots, formed on a barrel-shaped
mould and provided with a circular opening at the top
through which the charge was introduced and a similar opening
at the bottom through which the zinc vapour issued. When
required the pots were baked at a red heat and then placed in
the furnace. The method employed was an adaptation of the
existing process for brass manufacture.

At one time Champion had 31 furnaces producing zinc,
copper and brass, but later he suffered considerably from
commercial difficulties and in 1750 he brought a petition to

Parliament asking for an extension of his patent, complaining bitterly of the unfair war waged against him by Indian importers who had caused the price of the metal to fall from £260 to £48 per ton. The zinc imported from the Far East by the Indian traders was sold under the name 'spialter' or Indian tin, the name finally being modified to spelter, the term currently used for commercial slab zinc.

In Europe the Germans and Belgians conducted zinc distillation 'per ascensum' the zinc vapour being conducted from the upper part of the retort, condensation being effected in a tube or condenser attached to the retort. This process proved both thermally and metallurgically more efficient than the Champion method of distillation per descensum, and by the middle of the 19th century Bristol had lost the lead, Germany producing about 10,000 tons, Belgium 2,000, U.K. 2,500 and Poland 2,000, a total of 16,500 tons.

At the present time production is of the order of $3\frac{1}{4}$ million tons of which the following countries are the most important producers.

	Tons		Tons
Australia	134,000	Poland	193,000
Belgium	272,000	U.K.	83,000
Canada	260,000	U.S.A.	803,000
W. Germany	156,000	U.S.S.R.	380,000
Mexico	58,000	Elsewhere	876,000

According to the method employed in its recovery the metallurgy of zinc can be divided into two main branches. In the pyrometallurgical method the sintered zinc ore is mixed with powdered coal or coke and exposed to a high temperature in retorts made of refractory material. Zinc distils off and is condensed and collected. The main reaction concerned in the reduction is $ZnO + C = Zn + CO - 57$ K cals.

In reality the reduction takes place in two stages

$$ZnO + CO \leftrightharpoons Zn \text{ (vap)} + CO_2 - 18 \text{ K cals.}$$

$$CO_2 + C \leftrightharpoons 2\,CO \qquad\qquad - 39 \text{ K cals.}$$

The second reaction is slower than the first below 1100° C. and hence controls the rate of the reduction. Both reactions are highly endothermic.

Early furnaces were direct coal-fired and held a number of small retorts supported in a horizontal position and batch operated. In spite of many improvements in furnace construction and fuel economy, the size of the retorts has practically remained unchanged until the advent in 1930 of the continuous vertical retort process with large retorts mechanically charged and discharged.

The other branch of zinc metallurgy is an electrolytic process, the roasted zinc ore being leached with dilute sulphuric acid, the zinc solution then being purified and electrolysed to produce metallic zinc. The process, which was developed much later than the pyrometallurgical method, now produces about 50% of total zinc output.

The primary step in the production of zinc is the roasting of zinc blende which is the main economic ore. This involves heating the ore in a suitable furnace to convert the sulphide into zinc oxide with the production of sulphur dioxide

$$ZnS + 3O = ZnO + SO_2$$

The objective is to remove as much sulphur as possible, for retention of sulphur in the calcine causes the loss of an equivalent amount of zinc in subsequent processing. Many types of furnace have been designed and used. The first of these were the hand-rabbled reverberatory of the type employed in roasting lead and copper. These were succeeded by mechanically rabbled furnaces such as the Ropp straight-line furnace, the O'Hara, the Brown (both horseshoe and straight), the Merton superimposed hearth furnace, etc.

When the sulphur dioxide was to be utilized for the production of sulphuric acid a muffle-type furnace was employed to which the furnace combustion gases had no access, the ore being contained in a muffle chamber externally heated. The first plant built to utilize the sulphur gas from blende roasting was that at the Rhenania Chemical Works at Aachen in Germany in 1855. This was the invention of F. Hasenclever and consisted of three long (40 ft.) superimposed muffles enclosed in a firebrick furnace with a firegrate at one end. The combustion gases passed under the lowest muffle, then over the top muffle, the middle muffle deriving its heat from the oxidation of the blende induced by the heat from the top and bottom muffles. The ore was charged through a hopper at the top and was worked by hand-rabbling along the hearth to an opening at the extremity through which it dropped to the next muffle, being withdrawn from the lowest hearth through a door. The sulphur dioxide passed through all the muffles and out of an aperture in the roof of the top muffle to a flue from which it was conducted to the acid plant. The capacity of the furnace was about 5 tons of blende per 24 hours, the sulphur gases containing 6–7% SO_2. The consumption of fuel was high, being about 25% of the weight of the blende and the work of rabbling and moving forward of ore was arduous.

In the United States, with its high labour costs, the emphasis was on a mechanical type of muffle furnace and led to the evolution of the furnace introduced in 1881 by E. C. Hegeler of the Mathiesen and Hegeler Zinc Works, La Salle, Illinois. In this furnace the blende was rabbled mechanically by rakes drawn through the muffle by rods, the ore being rabbled every two hours or so. Although effecting good desulphurization the intermittent rabbling engendered hearth accretions which were a nuisance especially when treating leady ore, gas production was fluctuating and of poor quality and fuel consumption high. The introduction of the McDougal cylindrical multiple-hearth mechanical furnace with its compactness and more efficient operation and its subsequent improvement in the early part of this century by Herreshoff, Wedge, Skinner, etc., led to the gradual supersession of the straight line type of furnace. In turn the McDougal type of roaster for blende roasting has been replaced by the Dwight Lloyd sintering machine (page 60) in which desulphurization is effected in thin layers of ore by oxidation induced by internal combustion. It differs radically from all hearth-roasting methods in that oxidation is propagated throughout the entire mass of the ore and not as in hearth roasting wherein the oxidizing gases are in contact only with the surface of the ore which therefore has to be continuously stirred in order to expose fresh surfaces of ore to the oxidizing atmosphere. In addition external heat is required only for the primary ignition and not as in hearth roasting, where external heat is required for the entire operation. The advantages are such that sintering is now in almost universal use for blende roasting.

Sulphur dioxide derived from roasting forms the principal raw material for sulphuric acid; hence zinc plants tend to be located at centres where this chemical is used in large quantities. The availability of cheap sulphuric acid also offers a considerable inducement to the manufacture of other acid-consuming materials in conjunction with zinc such as ammonium sulphate, fertilizers, etc.

Smelting.　Present-day procedure is a modification developed

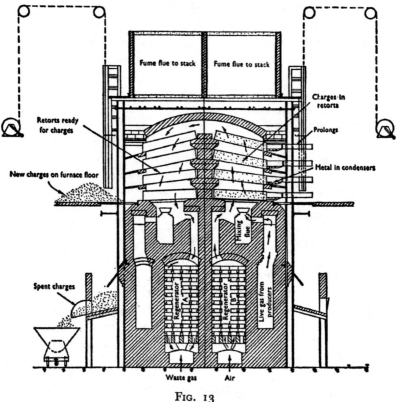

Fume flue to stack　Fume flue to stack

Charges in retorts

Retorts ready for charges

Prolongs

New charges on furnace floor

Metal in condensers

Mixing flue

Spent charges

Regenerator "A"

Regenerator "B"

Live gas from producers

Waste gas　Air

FIG. 13

ZINC HORIZONTAL RETORT FURNACE

by a combination of the best features of the Belgian and German (Silesian) methods, and consists in the distillation of roasted ore contained in small horizontal retorts suspended in rows in the combustion chamber of a furnace (Fig. 13) and so arranged that the flame circulates round them, the zinc which distils being condensed in individual condensers attached to the end of the retort. Whilst the last hundred years have seen improvements in the design and construction of the furnaces, the process of horizontal retorting in its essentials is much the same today.

The retorts are made of burnt clay finely ground and mixed with an approximately equal amount of raw fire clay to serve as binder. The mixture is thoroughly kneaded to remove occluded air and then made into retorts. These were formerly made by hand but now a hydraulic press is employed, the clay being first pressed into a cylindrical mass and then fed through a die in a press in which is positioned a mandrel so that the clay flows through the annular space formed between the die and mandrel. The retorts are then placed in vaults provided with steam coils and left to season for about two months. When required, they are annealed at a dull red heat. Commonly they are elliptical in cross-section, the long axis being about 11 in. and the short one 8 in., 4–5 ft. in length and holding about 120–150 lb. of charge. Their life is 35–40 days.

Condensers are made of clay and old retorts which have been crushed, the ground mixture being moulded bottle-shaped in a machine. Generally they are fitted when in service with sheet-iron canisters known as prolongs which project from the mouth of the condenser and serve the purpose of condensing any zinc escaping from the condenser, which otherwise would be lost to the atmosphere.

The furnace consists of an arched firebrick chamber, strongly braced by means of buckstays and tie rods and in which the retorts are arranged in 4 or 5 rows, one above the other, supported only at either end in order that they may be surrounded by flame. Formerly, a fireplace was situated at the bottom of the furnace chamber, but in modern practice producer gas is employed as fuel.

Regenerative checkerworks are usually placed underneath the furnace, the gas and air being preheated by entering the checkerworks, mixing in the flue above, the burning gas then passing up through the rows of retorts and over a bridge wall, descending between the retorts on the other side to the regenerative chamber below, preheating this as they pass through, and thence exhausting to the stack. At intervals of about 30 minutes, the direction of the currents of gas and air is automatically reversed. To conserve heat the furnaces are for the most part built in rows back to back.

Horizontal Retorting. The sintered ore is mixed with finely ground anthracite in the proportion of 30–45% by weight. This is 3–4 times the amount required by theory, the excess being required to form the extra reaction surface necessary to keep the carbon dioxide concentrate as low as possible ($C + 2CO_2 = 2CO$) for carbon dioxide reduces the yield of metallic zinc by oxidation of zinc vapour.

Charging of the retorts is done by scoop-shaped shovels, and compressed by blows with an iron rammer. When filled, the condensers are luted on with fireclay, the prolongs fixed and the temperature of the furnace gradually raised. Coal gas first issues burning at the mouth of the prolongs with a luminous flame which changes to bluish green as metal begins to condense in about 4–5 hours after charging and at a temperature of 1000° C. A temperature of about 1400° C. is maintained in the furnace, that in the retorts being 100–200° C. lower and in the condensers 450–550° C. The zinc which distils over and which remains molten in the condenser is usually ladled out 3 or 4 times during the distillation period into an iron pot or kettle (carried on an overhead rail in front of the furnaces) and cast into slabs. The first draw, i.e. the metal drawn during the first part of the distillation, is relatively pure (98·5–99·0%) and will meet many specifications without further treatment. Succeeding draws, however, are always higher in lead and iron. Cadmium, being a metal of lower boiling-point, is found mostly in the first draw. Wherever lead and iron are present the last draw gives a metal which is higher in lead and iron than the first draws. Condensation of part of the zinc vapour other than in metal form constitutes a serious loss of zinc. It occurs principally in the form of globules of zinc coated with from 5–15% oxide, the oxide coating causing segregation of the material. It is caused by the back reaction between zinc vapour and carbon dioxide ($Zn + CO_2 \rightarrow ZnO + CO$) and known technically as blue powder. The chief loss is due to retention of metal in the residue remaining from the distillation. It is due chiefly to metal present in the form of sulphide ascribed to imperfect elimination of sulphide during sintering, such sulphide not being reduced to metal during distillation. Other losses are due to absorption of zinc by retorts, escape of zinc vapour due to incomplete condensation and escape from cracked and broken retorts.

The direct extraction of zinc from the metals is only 65–70%, but owing to the recycling of such material as ladle skimmings, flow cleanings, blue powder, etc., recovery may reach 90%. Consumption of fuel in the regenerative furnaces is 1–1·5 tons of coal per ton of ore which compares to 4–5 tons in the old direct-fired furnace.

Although the metallurgy of zinc in the past hundred years has been marked by the development of the continuous vertical method, the reflux system of refining and the introduction of electro-winning methods, horizontal retorting remains much as it was with muscular metallurgy, dust and heat and fume still being well to the fore. The multiplicity of small retorts require a great deal of labour for charging and discharging. Mechanical charging and discharging machines have been devised, but their use has not proved popular; the machines are complicated and require a good deal of manœuvring to charge and discharge hundreds of small retorts. The retorts have to be of limited size, for the fireclay walls of the retorts being of low heat conductivity are not capable of sustaining a heavy charge at high temperature. A further factor limiting the length is the difficulty of shovelling the charge to the far end of the retort.

Vertical Retorting. The vertical retort process developed in 1929 by the New Jersey Zinc Co., Palmerton, Pa., overcame the above-mentioned difficulties and made practicable the smelting of zinc ore in continuous mechanical operations in large retorts. Increase in the size of the retorts was achieved by constructing them of individual bricks and placing them in a vertical position. This removed the weight strain from the walls of the retort and reduced the tendency to crack. A furnace chamber encloses each retort, heat being supplied by producer gas which enters at the top of the furnace chamber, the products of combustion passing out at the bottom to a recuperator and thence to a coker for coking the briquettes. The units, commonly measuring 30 ft. high, 7 ft. long and 1 ft. wide, are made of silicon carbide bricks which possess high heat conductivity and a low coefficient of expansion so that heat can readily pass through the retort walls. By forming the charge of briquettes that retain their form during the smelting operation, voids are provided through which the heat can be transferred from the retort walls to the charge by radiation and convection.

Operation. Sintered zinc ore and coal are crushed and mixed in the proportion 50–50, briquetted in a press and then hardened by coking at a temperature of 750–900° C. in a vertical oven through which the briquettes continually pass, the heat required for the oven being supplied by the exhaust gases from the vertical retorts. The coked briquettes discharge to steel buckets and are lifted by electric hoists and charged into hoppers at the top of the retorts. As the briquettes descend in the retorts, zinc vapour and gas pass upwards to an inclined conduit, thence to the condenser in which the molten zinc collects, the molten metal being tapped into ladles at intervals and cast into slabs. The spent briquettes sink to the bottom of the retort and are continually discharged by a screw conveyer closed by a water seal, the residue falling into cars for conveyance to the dump. Each retort is capable of producing 6 tons zinc per 24 hours, the overall recovery being approximately 92%.

Blast-Furnace Smelting. The vertical retort still does not overcome the disadvantage of indirect heating and the multiplicity of small units with high maintenance cost and upkeep. One method of avoiding the limitation of retort operation or the necessity to use electric power (i.e. electrolytic zinc deposition) is to utilize the economics of large unit blast-furnace practice. After 25 years' study this has been successfully developed (1957) by the Imperial Smelting Corporation at Bristol. Orthodox methods of recovering the metal from the furnace bottom having failed, efforts were concentrated on the task of condensing zinc in the metallic state from the gases leaving the furnace top. The problem entailed many difficulties, e.g.

1. To avoid reoxidation of zinc within the furnace the whole charge must be maintained above the zinc reoxidation temperature.

2. To avoid reoxidation during the passage from furnace to condenser, the gases must not fall below this reoxidation temperature.

3. To avoid reoxidation in the condenser, the gases must be rapidly cooled.

The furnace finally evolved was in some respects similar to that used in lead practice and consists of a rectangular brick-lined shaft with water-cooled tuyères and a water-jacketed bottom section. A mixture of zinc sinter and preheated coke is fed through the furnace top sealed to prevent oxidation of the zinc vapour by a double bell hopper. Preheated air is blown in through the tuyères, sufficient heat being liberated to reduce the zinc oxide and fuse the gangue material present to form slag.

Gases leave the top part of the furnace through a column of coke heated electrically to avoid reoxidation during the passage from furnace to condenser. The gases approximate 5% zinc, 10% CO_2 and 20% CO, and pass into two condensers, one on each side of the furnace. In each condenser there are three vertical-shaft motor-driven rotors dipping in a lead pool in the bottom of the condenser. The agitation of the rotors creates an intense shower of lead droplets which cool the gas instantaneously to a temperature of about 450° C., the lead/zinc mixture continuously flowing to a separator. The mutual solubility of the two metals being low, zinc separates as a top liquid layer, continuously overflows a weir, and is cast into slabs, the lead being pumped back to the condenser. Gases leaving the condenser are scrubbed with water to recover blue powder zinc and then burnt, the heat being utilized to preheat the air and coke used for combustion. The zinc produced assays Pb $1-1 \cdot 5\%$, Fe $0 \cdot 02-0 \cdot 07\%$, Cd $0 \cdot 07\%$. Slag, matte and lead bullion are periodically tapped from the bottom of the furnace, the matte and bullion being retained in an external settler, the slag overflowing to a car for dumping. An important feature of the process is its ability to recover lead values contained in the ore, simultaneously with zinc, the lead content of the raw materials being obtained from the furnace bottom as lead bullion.

Refining. Zinc was first used for making zinc and brass castings, for which purpose it need not be of high purity. Relatively impure zinc can also be mechanically worked as metal or alloy in the form of sheets. The products of the horizontal retort process satisfactorily met the needs of the time, until the beginning of the present century when the advantages of pure zinc with which to make brass, which can be mechanically worked at great speeds and stamped and drawn with greater facility, became evident. Zinc obtained by distillation or blast-furnace methods rarely exceeds 99% purity, lead, cadmium and iron being amongst the commonest impurities. The need for pure zinc was eventually met by the development of electrolytic methods in 1916 of producing zinc which established a grade of 99·9% zinc as ordinary commercial metal; later, refining by rectification was introduced. The earliest method of refining was, however, the process of liquation, the 'refined' zinc of the European smelter being made for many years in this manner.

Refining by Liquation. This method is based on the liquation of zinc from its impurities when it is kept at a temperature near its melting-point. At this temperature zinc will only retain about 1·0% of lead in solution, the excess settling to the hearth of the furnace alloyed with some zinc. The iron impurity also settles as a zinc-iron alloy forming a layer above the lead. The operation of liquation is carried out in a reverberatory furnace, the flame being kept short of air, so that a reducing atmosphere is maintained above the molten zinc in order to keep oxidation to a minimum. The lead 'bottoms' which contain up to 6% of zinc are allowed to accumulate and then periodically tapped from the furnace. The layer of ferruginous zinc is also generally allowed to accumulate until the furnace campaign is over when the zinc is all ladled out and the mushy layer removed with perforated ladles. It will contain 1% or more of iron and 2% of lead. It is not possible to refine to less than 0·8% lead or 0·02% Fe.

Refining by Distillation. Rectification as a means of separating and purifying liquids by redistillation, integrates in one operation multiple stages of fractional distillation and condensation. It is extensively employed in the chemical industry, the principle being first applied to the refining of zinc by the New Jersey Zinc Co. Palmerton, Pa., in 1933.

Zinc boils at 907° C., cadmium at 778° C., lead at 1720° C., and iron at about 2900° C., and the spread in boiling-points is sufficient for separation by rectification to take place. The conventional bubble tray type of fractionating column used for rectifying liquids proved unsuitable for the rectification of zinc, the heavy density of zinc being a disadvantage on account of the great resistance offered to the flow of vapour through the bubble trays. For this purpose a rectification column was used composed of a stack of silicon carbide trays (42 in. × 21 in. × 6 in.). In this type of column construction the liquid metal cascades down through the series of trays, the metallic vapour passing up in intimate contact with the descending liquid metal. The liquid metal and metallic vapour thus travel countercurrently through the column, the liquid metal effectively scrubbing the zinc vapour, ridding it of lead, iron and cadmium. Heat required for boiling is supplied at the lower end of the column, the function of the upper part of the column which is outside the furnace boiler being to induce a process of refluxing.

To avoid the build up of pressure due to the large density of zinc, the depth of metal in each tray is regulated by an overflow through an opening in its bottom to the tray below. The openings are staggered so that the overflow from each tray is caught on the one beneath. In order to permit contact of the maximum possible proportion of the heated walls with liquid zinc and at the same time provide adequate space for vapour passage, the central part of the tray bottom is raised, the liquid metal thus being contained in a trough extending around the periphery of the tray.

Two columns are provided, zinc distilling away from lead and iron in the first column, and cadmium being volatilized after condensation from high purity zinc in the second column.

Operation. Crude zinc is melted in a pot and fed to the first column about half-way down, addition of metal being regulated to a rate which will ensure that all the trays are full of metal and overflowing steadily to the trays beneath. Metal is vaporized in the lower part of the column, by heat from the furnace, ascends through the trays, being scrubbed free of iron and lead by the liquid metal and passes to the condenser. The feed to the column is maintained at a greater rate than the vapour output, the refluxed excess passing out from the bottom of the column via a molten metal seal to a run-off pot. This run-off product, which contains the bulk of the high boiling impurities such as lead and iron, can either be sold as low grade zinc or else recirculated after liquating out the lead and iron.

The cadmium column into which the lead and iron-free zinc flow from the condenser is similar in design to the lead column, except that the condensing capacity at the top is adjusted to the boiling capacity at the bottom, so that only a small proportion of feed containing the low boiling-point cadmium escapes at the top as vapour. The bulk of the feed freed from its low boiling-point impurity refluxes to the bottom of the column escaping to a pot from which it is cast into slabs averaging 99·99% Zn. Cadmium, which has been collected at the top of the column in a sheet-iron canister is refined into metal of 99·9% purity.

Electrolysis. Although the purification of many crude metals such as copper, nickel and lead by electrolysis is now firmly established, a similar process for the purification of crude zinc is not operated today. It was practised for a time in Silesia by G. Nahnsen in the 1880s. The metal produced assayed 99·9% zinc, 0·06% lead, 0·01% iron. The method is, however, not in use today, probably because of the success of electro-winning methods in producing metal of high purity direct from the ore.

The electrolytic zinc plant of today is largely based on patents issued to L. Létrange of Paris in 1881 and 1883. Ore was first ground and an effort was made to mix calamine and blende in such proportions that during roasting a maximum of zinc sulphate would be formed. The calcine was then leached with either water or sulphuric acid. Létrange does not specifically describe the purification of the zinc sulphate solution, but merely refers to purification by 'known means'. He electrolysed the solution using anodes of carbon and cathodes of zinc sheet, the cell being a wooden tank lined with sheets of lead or glass. He exhibited examples of his cathodes at the World's Fair in Chicago in 1893, but as little more was heard of him he apparently did not overcome the difficulties associated with the process. This was a matter of some surprise, for, with the exception of carbon anodes which are now known to affect electrolysis adversely and the use of zinc cathodes which are now displaced by aluminium sheets to obtain better stripping of the deposit, Létrange's method is practically that of modern practice. Only elaboration in detail and stringent purification of solution were needed for success.

Numerous patents were taken out in the next ten years, most of which dealt with sulphate solutions. C. Hoepfner, however, took out a patent in 1890 which dealt with a zinc chloride solution. By this process the ore was roasted with salt to convert it into zinc chloride, the zinc being leached out with water. The solution was then purified by addition of chloride of lime and calcium carbonate which precipitated iron and manganese, the other metals (copper, lead, etc.) being removed by the addition of zinc dust. The pure zinc chloride solution was then electrolysed using carbon anodes and zinc sheet cathodes, pure zinc being deposited at the cathode. Chlorine, which was liberated at the anodes, was utilized in the manufacture of bleaching powder by absorption in lime. A current density of 10 amps per sq. ft. with an E.M.F. of 3·3–3·8 volts was used. Hoepfner built two plants, one at Fuhrfurt-am-Lahn and the other in 1896 at the plant of Brunner Mond & Co., Winnington. This latter plant had an annual production of about 1,000 tons of high grade zinc and remained in operation until 1924. No commercial plant using chloride solution has, however, operated since then.

Whilst Hoepfner was busy in Germany, E. A. Ashcroft was developing a process in England. He proposed roasting zinc blende, leaching out the soluble zinc with ferric chloride, whereby zinc was taken into solution as chloride, the iron being precipitated as ferric hydroxide. This material was then electrolysed first in a series of cells provided with iron anodes and then in a second series of cells with carbon anodes. Zinc was deposited in the first series of cells with generation of ferrous iron which in the second series of cells was oxidized to ferric iron for use in the leaching step. Broken Hill Mine in Australia became interested in the process for the treatment of their mixed (lead and zinc) ore, the treatment of which at that time was one of the great metallurgical problems. Ashcroft in 1896 built a plant (costing £200,000) with a rated capacity of 10 tons of zinc per day at Cockle Creek, New South Wales, in 1898. The process, however, ran into so many difficulties that it was shut down soon after. The difficulties stemmed from inefficient roasting leading to formation of insoluble zinc ferrate, presence of gelatinous silica leading to difficulties in filtration and washing, poor purification practice and the presence of large quantities of cobalt causing re-solution of the deposited zinc.

The reason for failure of the earlier plants using sulphuric acid leaching was the lack of appreciation of the very high degree of purity of solution required for successful electrolysis. With the right kind of electrolyte, the electrolysis proceeds more or less smoothly. Ideas on what constitutes a pure solution have been revised by the passage of time. A statement 50 years ago that a fraction of one part per million of some impurity would seriously affect electro-deposition would have been received with some scepticism; yet now assays are regularly reported in parts per million.

With the outbreak of World War One, and the rapid rise in price offered for pure zinc (required for cartridge brass) research was intensified, and once it was established that a solution of high purity was the requisite for successful electrolysis, the success of the process was assured. The outcome of this research was that in 1916 two plants were erected.

That at Trail, British Columbia, was erected by the Consolidated Mining and Smelting Co. of Canada, at a rated capacity of 1,800 tons of cathode zinc per month. The other was erected at Great Falls, Montana, by the Anaconda Copper Mining Co. with a capacity of 3,000 tons per month. The success of these two plants was followed by the erection of electrolytic zinc plants in many other countries where cheap electric power was available, and this tendency was continued as can be shown:

Method	Proportion of World Zinc Production (1960)
Horizontal Retort	36%
Vertical Retort	7
Blast Furnace	3
Electro-thermic	3
Electrolytic	51

The method has for its object the production of a pure solution of zinc sulphate, followed by the deposition of the zinc from solution by electrolysis. To this end the process can be divided into four stages:

1. Roasting of raw ore to render soluble the zinc;
2. Leaching and purification;
3. Electrolysis;
4. Melting and casting the electro-deposited zinc.

1. *Roasting.* The raw material consisting of flotation concentrates containing 50–60% of zinc, 30–35% of sulphur, and 3–10% of iron is commonly roasted in the mechanical multi-hearth type of machine.

The charge is fed to the top drier hearth and is swept across by the rabbles and drops through a hole to the next hearth below and so on down the furnace. The material reaches a roasting temperature by the time it has travelled about a third of the height of the furnace; at the middle it is approximately 650° C. and at the discharge end 600° C. The roasted calcine is finally discharged at the periphery of the bottom hearth into a steel push conveyor and in travelling about 100 ft. is cooled from approximately 600° C. to about 100° C. Pulverized coal, gas or oil are the fuels used. Of the feed, 10–15% passes off as flue dust and is collected and returned to the roaster.

The main object of the roasting operation is to render soluble as much as possible of the zinc in the sulphuric acid electrolyte used for leaching. A dead roast is obtained only at an elevated temperature which in turn increases formation of ferrate, which is insoluble. Since any sulphide sulphur left in the calcine ties up approximately twice the weight of zinc, the most efficient roast demands a balance between the sulphur left in the calcine and ferrate produced in the calcine. In order to avoid the formation of the latter, temperatures in excess of 700–750° C. are avoided. The desirable type of calcine contains about 0·3–1% of sulphur as sulphide and 1–2% as sulphate with the iron thoroughly oxidized, complete volatilization of arsenic and chlorine having been achieved.

2. *Leaching and Purification.* In leaching, the common practice is a two-stage (acid and neutral) counter-current continuous process, although local conditions and character of material treated may cause some variations.

Essentially the process has two functions, (1) leaching and extracting the maximum amount of zinc.

$$(ZnO + H_2SO_4 = ZnSO_4 + H_2O)$$

and (2) the purifying of the resulting solution to yield a satis-factory electrolysis in the tank house.

Calcine is added to acid (spent electrolyte), the addition being so controlled that a large excess of calcine over that required for neutralization remains in the pulp. This effects precipitation of silica, alumina, iron, etc., and causes soluble silica to precipitate in the granular form, thereby aiding the subsequent filtration and washing. The leaching is carried out in a series of Pachuca tanks, acid being added to the first tank only and is so regulated that the available acid in the last Pachuca is of the order of

1·5 gm. per litre. If this strength is not adhered to, resolution of impurities which have been precipitated may take place.

The neutral overflow from the last Pachuca discharges to a series of thickeners, the underflow forming the feed to the acid leach. Here it is further agitated with acid, thickened, filtered and washed, the residue being delivered to the smelter for recovery of zinc, lead and any precious metals.

3. *Purification.* The overflow from the neutral thickeners free of iron and aluminium (which have been precipitated as hydroxides in the Pachucas) typically contains elements such as copper, cobalt and arsenic, which lower the hydrogen over-voltage, cause serious loss of current efficiency, some even when present to the extent of only one part per million, antimony, germanium and tellurium being outstanding examples. Cadmium up to 200 mg/l causes no loss of efficiency, but being co-precipitated with the zinc reduces its grade.

Cadmium and copper are removed by agitation with small amounts (1·5 g/l) of zinc dust and cobalt removed from the purified solution by addition of nitroso-beta-naphthol.

Assay of a typical purified solution is as follows:

Zn 120 g/l. Fe 0·015 g/l. SiO_2 0·10 g/l. Cd 5 m g/l.

Cu 0·1 mg/l. As and Sb 0·1 mg/l. Co 3·5 mg/l.

4. *Electrolysis.* The cells consist of lead or rubber-lined concrete tanks, anodes being composed of pure lead sheets ¼ in. thick, welded to lead-covered busbars. Cells are conveniently arranged in cascades of 6–8 cells with a drop of 5–6 in. between each cell of a cascade. Cathodes are of high grade aluminium sheet ⅛ in. thick, spacing being 3 in. centre to centre.

Cathodes numbering 20–30 and anodes 21–31 constitute a complete tank complement. Cathodes are frequently made with a small groove about ½ in. from the edges which makes a line of weakness in the deposited zinc sheet, and hence facilitates stripping.

Purified electrolyte flows from a storage tank to the head cell of each cascade, flowing from the upper cells to the lower, the overflow from the last cell passing to spent solution storage tanks for return to the leaching division for treatment of fresh calcine. Glue or gum arabic and beta-naphthol at the rate of

0·30 lb. and 0·20 lb. per ton of cathode zinc respectively are added to the feed in order to stabilize the solution and to reduce the formation of trees, i.e. irregular outgrowths of metal on the cathodes which in extreme cases cause short circuiting. Period of zinc deposition varies from 48–72 hours when the cathodes are lifted from the tanks by an overhead chain block and stripped by hand. The aluminium cathodes are then scrubbed, cleaned and replaced in the cells. One tank containing 30 cathodes with a deposition period of 72 hours yields nearly 1 ton of zinc. Cell temperature is an important part of tankhouse control for the effect of impurities on current efficiency is in direct proportion to rise in temperature. On the other hand the electrical conductivity increases with increase in temperature. Hence a balance has to be struck between these two factors, a temperature of 35° C. having been found to be satisfactory. During electrolysis, manganese (present in the ore and used to oxidize iron in the neutral leach) is oxidized at the anode and precipitated as manganese dioxide. Periodically groups of cells are cut out of the electrical circuit, solution pumped out and the manganese sludge removed and sent to the leaching plant.

Electrical data relevant to cell operation is given below:

Current load, amps.	10,000
Current density, amps per sq. ft.	30
Current efficiency %	92
Voltage per cell	4
Kw–Hr per pound of zinc deposition	1·6

Melting and Casting. Cathodes are commonly melted in reverberatory furnaces of 50–100 tons capacity at a temperature of 550° C. A distinctly reducing atmosphere is maintained in the furnace to avoid burning and oxidation of zinc. Dross consisting of a mixture of zinc oxide and metallic zinc is the invariable accompaniment of the melting, some 3–6% of the weight of cathodes charged to the furnace resulting in the formation of dross. Addition of ammonium chloride (2 lb. per ton of zinc) helps to separate the globules of zinc from the zinc oxide and causes the former to coalesce. The dross is raked out through the doors and taken to a dross furnace for further treatment.

Metal is tapped from the furnace into a casting pot, held by a chain block and cast into iron moulds which produce a slab weighing about 60 lb. The moulds are pivoted in order to facilitate emptying.

The zinc assays 99·97% Zn; 0·012% Pb; 0·001% Cu; 0·006% Cd; 0·001% Fe.

Low frequency electric induction furnaces with a capacity of 20 tons of molten zinc and a melting capacity of 3 tons per hour, are also in use for melting cathode zinc.

GALVANIZING

The familiar galvanized iron roofing sheet typifies the principal use of zinc, namely as a protective coating for iron and steel. Iron and steel surfaces will rust when exposed to weathering due to the combined action of oxygen and moisture. In order to resist such action coatings both metallic and non-metallic, are applied to prevent corrosion, and of these zinc is one of the most important, nearly half (1,400,000 tons) of the total world output being used annually for protecting iron and steel against rust. The protection afforded is due not so much to the coating effect but to the fact that zinc being electro-negative behaves as an anode and corrodes preferentially; if the coating is undermined or damaged it suffers sacrificially, the underlying iron or steel remaining unaffected.

Value of zinc as a rust-proof coating for iron and steel has long been appreciated, the earliest patent for galvanizing by means of immersion in the molten metal being taken out by Crawford in England in 1837 and in France by the chemist, M. Sorel. Incidentally corrugated sheet which was used in the ancient world to stiffen bronze was first adopted for iron roofing to increase the load carried by a given thickness by R. Walter of Rotherhithe a few years earlier (1828). Procedure in the early years of the 19th century consisted of immersing puddled iron sheet in a molten zinc bath contained in an iron pot heated with coal or coke. Upon attaining the temperature of the zinc bath the sheet was withdrawn. The coating on these early galvanized sheets was rough, heavy and relatively non-adherent. With the advent of the Bessemer and open-hearth steelmaking processes and improvements in the rolling procedure, sheet of greatly improved quality and uniformity was made available, and the production of a continuous firmly adherent zinc coating was made possible.

The production of a zinc coating involves the formation of a thin layer of intermetallic compound of iron and zinc alloy between the outer layer of pure zinc and the iron or steel basis metal. Since this alloy layer is somewhat brittle, its thickness is usually kept to a minimum and the various conditions involved in galvanizing are controlled with this end in view. Excessive bath temperature, prolonged immersion and slow removal of the article from the bath all favour thicker alloy layers. Addition of aluminium to the bath tends to the production of a minimum alloy layer and the addition of $0 \cdot 2\%$ in sheet galvanizing results in a thinner and more flexible coating.

After removal from the bath the coating begins to solidify in the form of crystals. These may vary in size from the very small form which give the surface a matte appearance to the large crystals which go to make up the 'spangle'. The size and appearance of the crystalline surface is also affected by small additions of tin and antimony to the galvanizing bath.

After dipping, the articles will carry $1 \cdot 8$ to $2 \cdot 2$ oz. of zinc per square foot of surface which after 'wiping' is reduced to a minimum of 1 oz./sq. ft.

Zinc coatings may be applied by methods other than hot dipping. Sherard Cowper-Cowles, an Englishman, in 1900 introduced the method of 'sherardizing', in which the articles to be coated are placed in a closed alloy steel box containing zinc dust powder. On rotation of the box in a furnace at approximately 350° C. a firm adhesive corrosion-resistant coat of zinc is built up.

Zinc also lends itself to electro-deposition. The process was first used on a commercial scale at the Langbein-Pfanhauser Werke in Germany. The electrolyte used was a solution of zinc sulphate at a current density of 50–100 amps per sq. ft. The anodes were zinc metal plates, the metal to be plated constituting the cathode. Although not so widely employed as hot-dip galvanizing, electro-coated sheets and wire are now finding an expanding market.

LEAD

The earliest furnaces for the production of lead consisted simply of a pile of stones placed around a fire and so arranged as to leave openings which served for the admission of air, charcoal being used as fuel. In the course of time an artificial blast was contrived by the provision of bellows worked by a water-wheel. From this humble beginning it was not a far step to the first real furnace known as an 'ore hearth' introduced early in the 18th century. The ore hearth is essentially a low fireplace surrounded by three walls, with one or more tuyères at the back. This furnace which at one time was extensively used in the U.K. and America is still firmly entrenched in certain parts of the world for the treatment of high-grade galena concentrates, a purpose for which it is well suited.

Other furnaces for the treatment of lead were based on the reverberatory furnace used in Flintshire, North Wales, which was apparently designed in the year 1698 by a physician named Wright. These reverberatory processes enjoyed a great vogue during the 18th and 19th centuries, but have not evolved as far as that of the copper reverberatory, for they were superseded by the blast furnace which has developed in much the same way as the copper blast furnace and is now the leading producer.

Development of the Ore Hearth. The ore hearth can be likened to a small rectangular blast furnace composed of a cast-iron trough or basin set in brickwork and surrounded on three sides by walls composed of iron blocks, the front being left open. In front, sloping down from the trough is a cast-iron work plate with a groove at the lower end which conducts the molten lead into an iron pot as it overflows from the hearth during smelting of the ore. Air for reduction is supplied by tuyères situated in the back wall.

The type of reaction used for the recovery of lead in the ore hearth is that known as air reduction and is based on the fact that when galena (PbS) is gently roasted with free access of air, a mixture of oxide and sulphate is first formed. On rise of temperature, lead sulphide reacts with the sulphate and oxide forming sulphur dioxide and lead.

Thus the reaction embraces not only the roast reactions

$$PbS + 3O = PbO + SO_2$$

$$PbS + 4O = PbSO_4$$

but also the reduction reaction as well, PbO and $PbSO_4$, reacting with the unoxidized PbS to liberate lead.

$$PbS + 2PbO = 3Pb + SO_2$$
$$PbS + PbSO_4 = 2Pb + 2SO_2$$

The two processes, roasting and reduction, go on simultaneously and in addition some PbO is reduced to metal by the fuel

$$2PbO + C = 2Pb + CO_2.$$

Ore hearths in the 19th century had a basin area of approximately 2 ft. square and 1 ft. deep, and held about 1 ton of lead, the work plate being 3 ft. wide and 2 ft. from front to back, having raised borders and sloped downwards about 5 in.

The back and sides of the furnace were composed of cast-iron blocks, the lower part of the back block being perforated for the admittance of a tuyère. A brick shaft over the hearth served to conduct away the fumes. After a short time of working, the hearth became very hot and in order to keep the temperature down, the back and sides were eventually provided with an internal open space or chest, the blast passing into this space thus providing some moderate cooling effect.

Method of operation was to shovel the ore on to burning coal and then operate the blast. After a suitable interval to allow the ore to become oxidized a rabble was plunged into the material in order to loosen and open up the caked mass. This was followed by raking a portion of the material on to the work plate, removing any slag, returning any half decomposed ore to the hearth. These operations were repeated every few minutes or so, the charge being manipulated with the rabble so that the material was subjected to the action of the blast. Molten lead seeped down through the charge so that the level of lead in the basin was raised, lead overflowing down the groove in the work plate to the lead pot.

Crusts were continually being formed, which if not broken up, prevented the uniform penetration of the blast. At the end of a 12-hour shift the hearth became too hot for working (the temperature attained by the charge being 900° C.) and smelting was suspended for a few hours. Each hearth smelted two tons of ore per shift yielding 1 ton of lead, two workmen being required per shift. Coal consumed was about 1–2 cwt. per ton of lead, the cost of smelting (in 1868) for 1 ton of ore being about 13/-.

In America early modifications to the ore hearth, included extension of the length of the hearth to 4 ft. and water-jacketing the back and two sides. Over the furnace a hood was placed through which a fan drew fumes and gases to a baghouse.

A radical improvement in the ore hearth was made at the plant of the St. Louis Smelting and Refining Co., Illinois, in 1914, by W. E. Newnham, who invented a mechanical rabbler; this was followed a few years later by the invention of the MacMichael mechanical shovel at the Federal Plant of the American Smelting and Refining Company.

With these two inventions the hard manual labour and exposure to fumes entailed by the stirring and poking were eliminated. In addition an increase in the size and capacity of the furnace was made possible by the extension of length of the hearth to 8 ft. Efficiency of the operation was increased and the amount of dust and fume reduced.

The equipment consists of a machine suspended from an overhead travelling carriage mounted on a track running parallel to the front of the hearth. The rabble and shovel are mounted on the machine and actuated through a series of gears, pinions and cam wheels driven by a motor. The carriage starts from the left-hand side of the hearth, a ratchet wheel moving the carriage forward about 4 in., the rabble arm then thrusting itself over the edge of the basin into the charge, raising and loosening the caked mass, after which it returns to the starting-point.

The shovel, which operates simultaneously side by side with the rabble is attached to a swing arm which is given a reciprocating movement and causes the shovel to execute a forward and backward stroke, the former throwing the material against the back of the hearth, a uniform penetration of blast and an even fire throughout being assured. On the back stroke the shovel is lifted clear of the material and is brought back in readiness for the next cycle.

The duration of each cycle is approximately 3 seconds, the entire traverse comprising 20 or so stages requiring approximately 1 minute. When the machine reaches the right side it automatically stops, allowing the operator to shovel the hearth slag into a car and mix in fresh ore and fuel. A lever is then pulled which causes a return of the machine (with the rabble and shovel retracted) to the left side of the hearth, ready for the next traverse.

The process gives rise to much dust and fume, approximately

35–40% of the ore mix charged, so that dust recovery equipment is essential.

Metallurgical data are as follows:

	Tons
Amount of ore treated per hearth per 24 hours	20
Lead content	14
Fuel used	2
Pig lead produced	10
Slag	4
Dust and fume produced	6

Recoveries %	
As Pig lead	70
As Lead and slag	11·4
As fume and dust	18·6

The higher the lead content of the feed, the greater the recovery. Silica should not exceed 2% and iron 4%, for these slag-forming constituents occlude both sulphide and metallic lead, prevent the elimination of sulphur and decrease lead extraction.

The ore hearth possesses the advantage that its first cost is much less than other types of furnace and it can be quickly stopped and started. At least 70% (with high grade 85% concontrate) of the lead is produced as bullion in one operation. On the other hand it is unsuitable for ore rich in silver on account of heavy volatilization losses.

Blast-Furnace Smelting. Hitherto the principles involved in the reduction of lead from its state of combination with sulphur have been considered with respect to reactions with air which is the basis of ore-hearth smelting. A further principle involves reduction of the ore by carbonaceous matter and is known as the 'roast reduction method', lead oxide or other oxidized compound being first prepared from sulphide ore by a separate roasting operation followed by a subsequent reduction with carbon in a blast furnace. As the method can treat economically all types of ore, it is now responsible for a majority of the world's output of lead.

The lead-blast furnace can be considered as a highly developed form of the ore hearth consisting of a vertical hollow shaft with a means for supplying an upward current of air to support combustion within a descending column of ore and fuel.

In the mid-19th century the furnaces were circular in cross-section, 4–5½ ft. at the tuyères and widened to a diameter of 7½ ft. at the mouth. Height from tuyères to mouth varied from 12 ft. to 20 ft. The furnace was provided with seven to eight tuyères, 2½ in. nozzle diameter, distributed around the circumference of the furnace, the amount of air injected being about 800 cu. ft. per minute at 6–8 oz. pressure. A forehearth extended 1 ft. or so in front of the furnace wall, the depth of the hearth below the tuyères being about 2 ft. The fuel was coke or charcoal with a consumption of about 25% of the weight of ore. The products from tapping were (a) lead bullion assaying about 97% Pb, (b) a matte assaying 20–25% lead, 19–22% sulphur, 25–35% iron which was roasted and resmelted; (c) an argentiferous speiss consisting chiefly of arsenide of iron and which was treated for recovery of its silver, and (d) a slag averaging 32% SiO_2; 40% FeO; 5% CaO and 3–6% lead. A lime flux was used, experience having shown that when so used the yield of lead was larger and the matte and slag poorer in lead.

Early types of furnace were built with firebrick linings which were so rapidly destroyed by the corrosion of the slags that maintenance of the linings became a full-time job, rendering the process largely uneconomic. It was not until B. Pilz at Freiberg in Germany had designed the water jacketed type of furnace in 1863 that the blast-furnace process took a step forward.

Circular furnaces cannot exceed a certain diameter, otherwise the air at the pressure employed cannot penetrate to the axis of the furnace and a cold central core of unsmelted charge is the result. If the air pressure is increased large losses of lead are incurred through excessive volatilization. Hence the adoption of a rectangular type of furnace which permitted the smelting area to be enlarged without increasing the distance between the tuyères.

A crude form of blast furnace with a rectangular horizontal section known as the Raschette was employed in the Harz district of Germany about 1864 for smelting lead ore. The furnace measured 3 ft. 2 in. × 6 ft. 7 in. at the tuyère level and

4 ft. 5 in. × 7 ft. at the top. Distance from tuyère to the top was 12–14 ft. Ten tuyères were used, five on each of the long sides. The furnace smelted 10 tons of roasted ore in 24 hours with a coke consumption of 32% by weight of coke. The labour of 2 smelters, 2 chargers, and 2 slag wheelers was required per 8-hour shift. A feature of interest was the water-cooling of the tuyères which protected them against the intense corrosion encountered in lead smelting.

The following advantages were claimed for smelting in the Raschette as compared to the ordinary circular blast furnace:

1. Production doubled.
2. Greater yield of lead.
3. Loss of lead in slags was reduced.
4. Production of less fume.
5. Longer duration of a smelting campaign.

In America, the length of the hearth was extended up to 96 in., the width, however, remaining at 40–60 in. depending on the blast pressure employed. This increase in the length of the hearth permitted a substantial increase in the amount of charge smelted.

About 1900 the length at the tuyère area had been extended to 140 in. long with a 16 ft. column and a 10 in. bosh, the walls above the bosh flared outward, the ratio of the area at the level of the charge to that of the area at the tuyère level being in the region of $2\frac{1}{2}$ to 1. This upward enlargement of the furnace shaft slowed up and expanded the ascending gases giving them more opportunity to spread evenly through the descending charge and thus effect a more efficient reduction, at the same time lessening the velocity and thus decreasing the amount of dust carried on to the flues.

At first the water-jackets which were composed of iron or steel sheet were confined to the tuyère area (Fig. 14) resting on the brickwork of the crucible and extended upward 4–6 ft., the shaft above being composed of the usual firebrick.

With the enlargement in capacity of the furnace and consequent increase in blast pressure, a second tier of water-jackets was added, so that the modern furnace has the entire shaft from the crucible top to the charging floor water-cooled. A large amount of furnace heat (10–15%) is absorbed by the cooling water, but this is more than offset by the protection afforded the furnace.

Another improvement effected in the 1880s was the invention of the siphon lead tap at Eureka by A. Ahrents. It consists of a sloping channel about 8 in. square leading from the bottom of the crucible to a basin on the outside of the furnace situated in the middle of one of the long sides of the furnace. It permits the

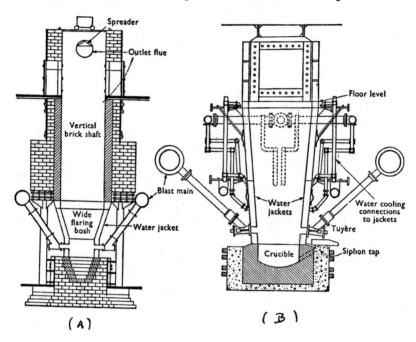

FIG. 14

LEAD BLAST FURNACES

(A) Dating about 1900 showing wide flaring bosh and vertical brick shaft
(B) Modern furnace showing bosh line of water-jackets in same straight line

maintenance of a crucible full of molten lead which rising in the channel fills the basin outside. Formerly, the crucible was periodically drained of its lead contents which created fluctuation of temperature leading to inefficient working and bad slag control. The siphon lead tap by permitting the maintenance of a bath of hot lead inside the furnace, smoothed out irregularities and led to better working efficiency.

The following table gives the principal dimensions of blast furnaces in use over the past hundred years:

Period	Dimension at tuyères (in.)	Smelting Column (ft.)	Blast Pressure oz./sq. in.	Tons smelted per day	Remarks
1860	48 — 60	6	12	7–10	Circular
1890	45 — 96	14	14	100	Rectangular
1900	45 × 140	16	24	150	,,
1920	48 × 160	18	26	210	,,
1940	48 × 200	20	36	500	,,
1960	60 × 280	20	38	580	,,

Recently (1950) increased capacity has been obtained by a new design. At the Broken Hill Smelter, Port Pirie, S. Australia, the lower portion of the lead furnace has the usual width of 5 ft. but 6 ft. above the tuyères the furnace is widened to 10 ft. and a second row of tuyères is introduced. A control core of unsmelted charge passes through to the lower zone where the smelting rate is so accelerated by the molten material seeping down from above that furnace capacity has been practically doubled.

Preparation of Feed. In order to render possible the reduction of lead ore in the blast furnace the sulphur has first to be eliminated. Reduction in the amount of arsenic is also desirable as this reduces the amount of speiss produced on smelting. The chief reactions that occur in roasting lead ore are the following:

$$PbS + 3O = PbO + SO_2$$
$$2PbS + 7O = PbO + PbSO_4 + SO_2$$

any copper and zinc sulphide present are similarly oxidized

$$3CuS + 9O = Cu_2O + CuSO_4 + 2SO_2$$
$$2ZnS + 7O = ZnO + ZnSO_4 + SO_2$$

Roasting Furnaces. The practice of roasting lead ore has gone through much the same evolution as that of copper roasting. Reverberatory furnaces of various types were formerly employed in roasting being in general use until the close of the 19th century. Beginning with the simple hand-rabbled furnace described on page 57, the arduous labour of hand-rabbling was replaced by iron rabbles dragged by chain along the reverberatory hearth. In the Brown 'horseshoe' furnace, so named because of its shape, the rabble arms carrying the rabbles were supported at the ends by small carriages drawn by endless steel ropes running in narrow chambers along the sides of the furnace, the chambers thus affording some protection to the rope from the heat and gases. Then came the Ropp furnace in the form of a reverberatory 105 ft. long, the rabbles moving on an endless track which continued outside and under the furnace so that the rabbles, after passing through the furnace, were drawn along the exterior track thus allowing time to cool before re-entering the furnace. The Bruckner cylinder which consisted of a refractory-lined steel cylinder revolving on friction rollers was popular at one time. Its chief disadvantage, however, was the production of much dust and semi-fused material which resulted in imperfect roasting.

Details of the above furnaces are given below:

	Hearth area (ft.)	Coal Consumption %	Output 24 hours
Hand-rabbled	50 × 12	25–30	5–15
Brown	105 × 12	16	25
Ropp	100 × 14	10	35–50
Bruckner	7 ft. diam.	12	8

During this period, the upright circular mechanical multi-hearth roaster came into use, and eventually displaced the reverberatory and other types of roaster. The mechanical roaster, however, suffered from the disadvantage of production of a large amount of dust which necessitated extensive equipment for its recovery. Further, the low melting-point of galena caused accretions to build up on to the hearth with consequent damage to the rabbles. For these and other reasons most of the mechanical roasters have now had to give way to sintering processes. Of these the first to be adopted was the Huntington–Heberlein method which was originated in Italy in 1896 and was superseded by the Dwight-Lloyd machine in 1905. The first D-L machine to sinter lead ores commercially was installed at Perth-Amboy, New Jersey, by the American Smelting and Refining Co., U.S.A., in 1907.

Sintering and the equipment used have been described on page 59 and it remains only to deal with the subject as applied to lead ores.

Sintering when first attempted on lead sulphide ore ran into difficulties for the high temperature developed not only fused the sulphide into matte, but also the slag-forming constituents coated the sulphide particles, thereby preventing any further desulphurization. The obvious solution was to absorb as much heat as possible by the incorporation of a dilutent. Water absorbs heat in evaporating and limestone on decomposition, and hence the incorporation of these materials in the feed lowered the temperature enough for proper desulphurization. It was, however, found difficult to get the sulphur down below 4% and this was not enough to avoid the formation of matte in the blast furnace. It had previously been found that if the ore had been given a pre-roast in a mechanical type of roaster, good results were achieved in sintering and hence the remedy was found in double sintering. The ore was first passed over the machine as rapidly as possible with light ignition, no effort being made to make good sinter, but simply to effect a preliminary pre-roast. The product was then crushed and run over a second time. This resulted in good sinter provided that a certain minimum of sulphur was present during the second sintering. If too low in sulphur this could be attained by addition of a small amount of raw ore.

A variation of this method employed in small plants possessing only one sintering machine is to incorporate with the fresh charge a certain amount—say 50% of crushed sinter. The result of this is a lower initial sulphur, thereby moderating the action.

Blast-Furnace Operation. The charge consists of sinter, fuel and miscellaneous material such as baghouse dust, scrap iron and refinery by-products, a typical charge being as follows— for comparison a charge of sixty years ago is also given:

	Today	1900
Sinter, %	90	—
Roasted Ore	—	20
Unroasted Ore	—	50
Limestone	1	18
Iron Ore and scale	3	8
Refinery dross, baghouse dust, etc.	6	4
Coke, %	10	15

The most noticeable feature is the amount of sinter incorporated in the present-day charge, and this is one of the main factors in securing the remarkable increase in output which has taken place. In the main it is a means of improving the physical nature of the finely divided lead flotation concentrate, but a substantial benefit also results from the reduction in sulphur, moisture and carbon dioxide. An economy in coke consumption also results.

With the increase in the size of the furnace, the transportation of the charge formerly carried in wheelbarrows is now usually effected by means of V bottom cars drawn by an electric locomotive, each ingredient of the charge being weighed. The charge from the cars is dumped on to inclined feed plates exterior to the furnace, sliding doors in the hood giving access to the interior of the furnace.

The charge as it descends in the furnace is subject to a gradually increasing temperature which reaches a maximum in the tuyère area of about $1200°$ C. which in conjunction with the reducing agent transforms it into bullion, matte, speiss, slag and gas.

The chief reducing agent is carbon monoxide formed by the action of the blast on coke:—

$$C + O_2 = CO_2$$

which in presence of more coke is reduced to carbon monoxide

$$CO_2 + C = 2\ CO$$

This reacts at a red heat with the lead oxide in the sinter, abstracting oxygen and liberating lead

$$PbO + CO = Pb + CO_2$$

Carbon is also a reducing agent

$$2PbO + C = 2Pb + CO_2$$

but the rate of reduction with solid carbon is relatively slow.

Reduction of iron ore and scale to FeO takes place by CO

$$Fe_2O_3 + CO = 2FeO + CO_2$$

the ferrous oxide combining with SiO_2 of the ore gangue to form slag. Part of the ferrous oxide is decomposed by C to metallic iron.

$$FeO + C = Fe + CO$$

which acts on any undecomposed sulphide to produce Pb

$$PbS + Fe = FeS + Pb$$

Dissociation of limestone

$$CaCO_3 = CaO + CO_2$$

commences at about 700° C., the CaO combining with SiO_2 in the gangue. Any copper in the charge will combine with sulphur if present forming matte, otherwise it will pass into the lead bullion. Arsenic will in part volatilize and in part combine with iron to form speiss.

These reactions are complete about two-thirds of the way down the furnace. Lead reduced in the upper part of the furnace seeps down through the charge collecting any gold and silver which may be present and joins the slag in the crucible.

Gases pass from the top of the furnace at a temperature varying from 150° C. up to 500° C. and contain considerable amounts of flue dust composed of fine particles of ore and fuel and fume composed of lead sulphate and oxide. Dust and fume are collected in baghouses or Cottrell precipitators, and incorporated in a charge to the D-L sintering machines.

The products of smelting are:

1. Lead bullion, which flows continually via the siphon tap from the lead well, being caught in a lead pot and transferred to the refinery for purification and recovery of its gold and silver content.

2. Matte. With increasing sulphur elimination in sintering, matte formation has decreased to such an extent as to be almost non-existent. As a consequence any copper is forced to go with the lead. Speiss likewise in recent times has virtually disappeared.

3. Slag. Slag is periodically tapped at intervals from a slag notch and flows into a settler, overflowing to a slag pot and is hauled away to the slag dump. It assays 20–40% SiO_2; 30–40% FeO; 5–20% CaO and $1 \cdot 0$–$2 \cdot 0$% Pb.

4. Fume and Dust. Gases arising from the furnace are combined with the gas from the sintering plant and travel through flues in which they drop most of their dust content and then to special collectors for recovery of fume.

Refining. Lead as received from the blast furnace contains up to 4% impurities as per a typical assay below:

Cu	1·0%	As	0·5%
Sb	2·2%	Bi	0·01%
Fe	0·3%	S	0·20%
Ag	110 oz./ton	Au	0·06 oz./ton

These impurities render lead hard, interfering with its malleability required for rolling out into sheet and forming into pipe, and their elimination is an economic necessity. In addition sufficient silver is usually present to render its recovery a profitable undertaking. Two methods are available for refining, namely fire methods and electrolysis.

Fire Refining. Preliminary refining operations remove those impurities from lead which render it hard, hence the operation is known as 'softening'. These impurities are mainly copper, antimony, arsenic and tin.

The preliminary operation takes place in oil-fired semi-spherical steel vessels known as kettles. Formerly the whole operation was carried out in reverberatory furnaces, but nowadays the tendency is for a preliminary decopperization drossing to be performed in kettles, thus reducing the excessive corrosion of the reverberatory furnace walls by the metallic oxides formed in the operation. After the softening operation desilverization takes place, four methods being available, (a) cupellation; (b) Pattinson's process; (c) Parkes' process and (d) electrolysis.

Cupellation was formerly in use for the extraction of silver long before the invention of other processes, but it has now practically ceased to exist as an independent process. It is, however, still used for extracting silver from the rich lead bullion formed in both the Pattinson and Parkes processes.

Preliminary drossing. The hot molten bullion from the blast furnace is received in a large (100 ton) kettle and agitated by means of a motor-driven impeller until cool. A copper dross rises to the surface which is removed, finely divided sulphur at a rate of about 2 lb. per ton of lead being then introduced to the bath, stirring being resumed, the temperature being kept as close as possible to the freezing-point (327° C.) of the bullion.

A sulphide of copper dross forms which is removed, leaving the bullion substantially free (0·005%) from copper. The copper dross is smelted in a small reverberatory with addition of a lime-silica flux at a temperature of 900–1000° C., a copper matte and speiss being produced for shipment to a copper smelter.

Softening. The drossed bullion is pumped to a water-jacketed reverberatory furnace, the temperature raised to 750° C. and the bath agitated by air blown through steel pipes. Tin is the first to be oxidized, its presence being denoted by the formation of a yellow-coloured stannate of lead. On removal of this slag, arsenate and antimoniate of lead are formed as a leady slag. Agitation is continued until the antimony content is less than 0·01%, when the slag is removed and the softened bullion transferred to the desilverization kettles. The softener slags containing up to 12% antimony are melted in a small reverberatory and reduced by addition of powdered coal. Two products result, an enriched antimonial slag containing 20–25% antimony which is treated for elimination of arsenic and tin and sold as antimonial lead. The other product a hard lead is either sold as such or re-treated.

Desilverization, Parkes Process. When zinc and lead are melted together and the molten mixture is allowed to cool, practically complete separation of the two metals occurs, the zinc on account of its slight solubility in lead and higher melting-point solidifying first and forming an upper layer which may be removed as a crust from the liquid lead underneath. A. Parkes (1813–90) of Birmingham, found that in the case of argentiferous lead, the silver becomes concentrated in the crust of zinc, and hence this greater affinity of silver for zinc than for lead constituted a basis for separation of silver from lead.

The process is carried out by pumping the softened lead to a kettle, any dross formed being skimmed off, and heating to above the melting-point of zinc (410° C.). At the melting-point of lead, zinc dissolves to the extent of 0·6%, and hence this amount of zinc has first to be added to saturate the lead before any silver will be surrendered. After that amount of zinc has been added, additional zinc must be provided to combine with the silver, the amount varying with the quantity of silver present, and the purity of the lead. Very roughly 15 lb. of zinc is required per ton of lead.

Usually two additions of zinc are required for desilverization of bullion. After the first addition the molten lead is thoroughly stirred by means of a mechanical stirring machine and after about one hour the stirring is stopped. On allowing the lead to cool zinc silver crusts separate from solution, rise to the surface and are removed. The lead is then reheated, a second addition of zinc made and the crusts which form on cooling are removed in the same manner. The time required for the operation is about 12–18 hours. On a 100-ton charge and a 200 oz. bullion, the first crust would weigh about 8,000 lb. and the second 20,000 lb., the desilverized lead assaying 0·1 oz. of silver per ton and 0·6% of zinc.

Dezincing Operation. Removal of zinc remaining in the lead after desilverizing is based on the preferential reaction of chlorine with zinc as compared to lead. The process commonly practised at the time of Parkes consisted of addition of lead chloride to the amount of 3½% to the molten lead. On stirring, the zinc is converted into chloride with separation of an equivalent proportion of metallic lead from that salt. The zinc chloride rises to the surface of the lead and is removed leaving a lead containing 0·0006% zinc. Alternatively, the lead was melted in a reverberatory furnace with admission of air, oxidation of zinc taking place, the leady zinc oxide slag being removed by skimming.

In 1928 a process of dezincing with gaseous chlorine was developed by J. O. Betterton of the American Smelting and Refining Co. The process consists of circulating the molten lead contained in a kettle at a temperature of 400° C. by means of a submerged pump through a closed reaction chamber, gaseous chlorine entering the chamber with the lead. Molten zinc chloride is produced and is discharged together with the lead from the reaction chamber and floats on top of the lead in the kettle. The zinc chloride is removed from the surface of the lead by skimming. Time taken for dezincing a 100-ton charge from 0·6 to 0·005% is 4 to 5 hours. The by-product zinc chloride is treated for removal of lead chloride and oxychloride and sold.

About 1945 a further method for elimination of zinc was

developed by the St. Joseph Lead Co., Josephtown, Pa. A large steel bell-shaped apparatus equipped with stirrer and attached to a vacuum pump is submerged in the molten lead. Temperature of the metal is raised to 540° C. and the vacuum pump started, zinc distilling from the lead and condensing on the interior surface of the bell, from which it is periodically removed and used for further desilverization.

Recovery of Gold and Silver. The zinc-silver crusts from the desilverization process, after removal of excess lead in a press, assays 20–60% zinc, 30–40% lead, and up to 25% silver, with varying amounts of gold. The first step, the removal of zinc, is performed by distillation in the manner originated by Parkes. Many other methods for the preliminary treatment of the crusts have been attempted in the past, such as (*a*) smelting of crusts in a blast furnace with coke as fuel, the silver accompanying the lead which is tapped off and refined by cupellation; (*b*) dissolving out the zinc by acids; and (*c*) direct cupellation. All these methods fail to effect recovery of zinc and the distillation method remains in universal use. The introduction by E. Balbach (1839–1910) of the graphite retort in place of fireclay has been perhaps the greatest improvement on the original method.

The crust is charged to the retort, a refractory-lined steel condenser being swung over the retort mouth and luted into position with clay. Zinc distils over and is periodically tapped to moulds giving slabs of zinc for return to the desilverizing kettle. The bullion is cast into bars for subsequent cupelling. The usual charge is from 1,000–3,000 lb., the distillation lasting from 12–20 hours.

Cupellation. This silver recovery process was mentioned in the Old Testament,* and is still in extensive use today. The principle on which cupellation is based is that when molten argentiferous lead is subjected to a blast of air, the lead oxidizes to litharge (PbO) which is easily fusible and acts as a solvent for other base metal oxides which may be present, such as Cu_2O and ZnO. The fluid mass of litharge is continuously run off leaving behind silver and any gold which remain unaffected.

The cupellation furnace is of the reverberatory type, com-

* Jeremiah vi. 29, 30.

monly measuring 9 ft. × 3 ft. The hearth, which is known as the 'test', takes the form of a shallow basin made of bone ash, dolomite or cement moistened with water and moulded by pressure on the top of an iron trough. As it is subject to severe corrosion from the molten litharge, frequent repairs are necessary and hence it is sometimes constructed independently of the rest of the furnace, resting on a small bogey, the whole unit being capable of removal and replacement by a fresh one. Air is supplied through tuyères at the back of the furnace, the furnaces usually being oil- or gas-fired. In operation the test is heated to redness and the bullion charged. When molten the air blast is turned on, oxidation beginning as a result of the impingement of air on the surface of the metal. The litharge formed runs out through a notch in the front of the test into moulds placed underneath the furnace. When the charge in the test has been concentrated to the required degree, the blast is stopped and the contents poured into moulds. It is usual to conduct the operation in two stages, a preliminary concentration taking place in one furnace up to about 50% silver, the metal from this stage being treated up to 99% silver in a further furnace.

The silver product obtained by cupellation contains any gold that may have been present in the bullion, and a finishing treatment for recovery of gold is necessary.

One of the earliest separations of gold from silver and base metals was effected by nitric acid which because of cheapness was replaced about 1840 by sulphuric acid parting. This in its turn was replaced early in the 20th century by electrolytic methods which are discussed in detail on page 288.

Pattinson Process. In 1829 an assayer named H. L. Pattinson (1796–1858) noticed that when lead is cooled to its freezing-point, crystals of lead separated out, which on assaying were found to be much poorer in silver than the still liquid lead. He obtained a patent in 1833 for his discovery. The process is usually operated in one of two systems, the system of 'thirds' or the system of 'eighths'. In the former, two-thirds of the molten lead in a kettle is removed as crystals, one-third remaining in the pot. The lead remaining will be approximately twice as rich in silver and the crystals one-half as rich as the original lead. In the second and less common system seven-eighths of the lead is removed as crystals and one-eighth left in the pot which will be about six times as rich as the original lead.

The 'thirds' system is conducted in a series of kettles, arranged in a row. The drossed lead bullion is allowed to cool and the primary crusts which form are stirred down into the lead to be remelted, the bath being continuously stirred in order to ensure uniform cooling. Crystals which now form are removed, transferred to the next kettle on the right, two-thirds of the lead being removed in this manner. The remaining liquid lead is then ladled into the pot on the left. The lead in the pots on the right and left are then successively treated in the same manner, the kettles on the right decreasing in silver value down to $0 \cdot 3$–$0 \cdot 5$ oz. per ton, the kettles on the left increase in silver until it reaches a figure of 500 oz. per ton, the maximum beyond which the enrichment of the lead cannot proceed. The material then goes to cupellation for separation of silver. Gold follows the silver, and of the base metals bismuth and antimony likewise follow the silver, and as a consequence the lead obtained is very soft and pure.

A comparison of the Parkes and Pattinson processes shows that the advantages of the former are:

1. Much less labour is required.

2. A much richer lead is produced for cupellation, 2,000–4,000 oz. silver per ton, as against 500–600 by Pattinson's process.

3. Produces a market lead lower in silver content, 4–6 dwt. as against 9–12 dwt.

4. Recovers all gold. On the other hand Pattinson's method produces a very pure lead, bismuth is removed and the equipment required is very simple.

Electrolytic Refining. The first attempt to refine lead bullion electrolytically was made in 1884 by Prof. N. S. Keith in America. He employed an electrolyte consisting of 180 gm of sodium acetate per litre in which was dissolved 20 gm lead sulphate. Anodes were cast of the base lead bullion to be refined measuring 24 in. $\times$ 6 in. $\times$ $\frac{1}{8}$ in. and weighing 8 lb. These were enclosed in muslin bags to collect the slime and prevent it dropping off to the bottom of the tank with the refined lead crystals falling from the brass cathodes. A plant was built at

Rome, New York, composed of 30 concrete tanks circular in shape, 6 ft. in diameter, the anodes being hung from a frame which rotated and carried scrapers that removed the deposited lead from the cathodes. Current used was 2,000 amps at 10 volts, the rated output being 3 tons of refined lead per day.

The Tommasi process (1896) employed a double acetate of lead and potassium as electrolyte. Cathodes consisted of a series of bronze discs mounted on a shaft just above the tanks which rotated between the anodes, the spongy lead deposits being scraped off mechanically as the cathodes left the electrolyte.

Neither of the above processes achieved much success on a commercial scale. When only low current density was employed deposits were reasonably compact, but as soon as the current density was increased in order to secure an economic output per tank, irregular outgrowths of metal would take place eventually resulting in short circuiting. Moreover, the solutions employed were so dilute that to effect a reasonable throughput a very large number of tanks had to be employed. The chief difficulty —incoherence of the cathodic deposit—was overcome in 1901 by A. G. Betts, who employed a lead fluosilicate electrolyte. This substance possesses a high electrical conductivity and is free from secondary decomposition, there being no polarization from formation of lead peroxide on the anode, as happens when acetate solutions are employed. Betts also found that addition of small amounts of glue or gelatine to the electrolyte favoured the production of a solid lead cathodic deposit. The first plant operating the Betts process went into production in 1903 at Trail, British Columbia, and it is now used by many other plants where low power rates prevail.

The electrolyte used is a solution of lead fluosilicate and hydrofluosilicic acid containing 7–10% lead, 12–15% total H_2SiF_6 and 9% free H_2SiF_6 prepared by dissolving litharge in H_2SiF_6. Cells are of reinforced concrete lined with asphalt, the electrolyte at a temperature of 40–50° C. being circulated through the cells at a rate of 3–4 gallons per minute.

Cathode starting sheets are cast from refined lead by ladling the molten metal over an inclined cast-iron plate. Anodes, composed of the lead bullion being refined, are cast into plates.

It is a characteristic of the electrolytic lead process that practically all the impurities commonly associated with lead bullion, i.e. Ag, Sb, As, Sn, Bi, are more 'noble' than lead, and accordingly remain undissolved at the anode during electrolysis in the form of a metallic slime layer or blanket which retains the original shape of the anode.

This tendency is encouraged as much as possible, for if the slime sloughs off into the electrolyte it is liable to be mechanically carried across and contaminate the cathodic deposit.

As solution of the anode takes place, the current must be carried by the electrolyte in the pores of the slime blanket remaining on the face of the anode. Within these pores the concentration of free acid becomes less and conversely the lead fluosilicate increases, leading to increased resistance to the passage of current with a consequent voltage drop. If this voltage drop attains a value equal to or more than the difference between the solution potentials of an impurity present in the slime and that of lead, such impurity will be dissolved and transferred to the cathode. Under these conditions, the blanket of slime will disintegrate, permitting silver and additional quantities of slime to be mechanically transferred to the cathode.

Generally the voltage drop across the cell increases gradually from 0·35 volt to 0·70 volt. On attainment of this latter value the anodes are removed from the cells, washed free of slime and transported to the scrap melting pots. Cathodes are removed from the tanks, washed free of electrolyte and any slime and charged to the refined lead kettle for melting and casting as refined lead.

Slime is periodically collected, washed and dried, and transferred to a reverberatory, the base metals being slagged off by oxidation leaving a gold-silver alloy which is cast into anodes for electrolytic parting.

Relevant electrolytic tank data is as follows:

Cathode current, amps/sq. ft.	15–17
Power	5,000 A at 110 V
Current Efficiency %	90–93
Anodes. Size (in.)	$36 \times 26 \times 1$
Anodes. Weight (lb.)	300–400
Cathodes. Size (in.)	$36 \times 26 \times \frac{1}{32}$
Cathodes. Weight initial (lb.)	12
Cathodes. Weight final (lb.)	150

Tanks. Size (in.)   110 × 32 × 48

Number of cathodes, anodes   25, 24

Anode slime %   Cu 1–2 Bi 3–30

               Sb 24–34 Pb 15–30

               As 7–13 Ag 300–3,000 oz. per ton.

Principal producers of lead at the present time are the United States, Mexico, Australia and Canada. In addition a very considerable amount of scrap lead is reclaimed annually to make the world consumption of pig lead about $2\frac{1}{2}$ million tons a year.

Production of Lead for 1960 (Tons)

Australia	272,069
Canada	160,079
Germany	162,772
Mexico	205,263
U.S.S.R.	350,000
U.S.A.	382,436
United Kingdom	1,224
Other countries (including Peru, Burma and Yugoslavia)	996,157
	2,530,000

In former times the chief use for lead was for roofing, sheathing of ships' bottoms as a protection against corrosion, and as a lining for sulphuric acid chambers. For these purposes the sheet lead required was obtained by casting over a rectangular table slightly inclined and having a raised border except at the lower end. The table was covered with a thin layer of fine sand. At the higher end of the table was a trough so arranged that by means of a lever and chain it could be raised and tilted in such a manner that when filled with molten lead, the metal would flow in a broad sheet over the table. Variation in the thickness of the lead could be attained by extra moistening of sand and by casting the lead at a lower temperature. The Chinese prepared their well-known tea lead by casting over a large flat stone and then pressing it flat into a thin plate by means of another flat stone. The lead sheet was then removed and a further quantity of lead poured, pressed, etc., the process being carried on with extreme rapidity.

Cast sheet lead is not now produced to any great extent. Old cast sheet lead from church roofs damaged by mechanical agencies is sometimes stripped, melted down, recast and replaced.

The process of rolling (or as it was originally termed, milling) into sheet by passage of metal between two rollers revolving in opposite directions appears to have been first practised in England in 1670 when a company was formed for the purpose of manufacturing lead sheet for the preservation of ships' bottoms.

Although not in extensive use today for ships' bottoms and roof covering, lead is used for gutters, weatherings to chimneys, roof flashings, etc. Lead piping, formerly produced by bending sheet lead over a mandrel and joining the overlapping edges by lead burning, is now produced by extrusion from hydraulic presses. Lead sheathing for telephone and power cables—which is the largest outlet for lead in Great Britain—is likewise produced by the process of extrusion.

NICKEL

Although nickel is a common constituent of the earth's crust and more widely distributed even than copper, it has so far been found in workable deposits only in Canada, New Caledonia, Finland, Russia, Cuba and a few other areas. By far the most important of these deposits is in the Sudbury region of Ontario, Canada, some 90% of the world's supply having been derived from this area. It is there that the methods used for its extraction have been largely developed.

The presence of ore deposits in this region was known as early as 1856 but owing to lack of communications and inaccessibility little interest was shown. It was not until 1883 when an extensive ore body was revealed during the driving of a cutting for the construction of the Canadian Pacific Railway that interest in the mineral riches of the Sudbury area was awakened.

The first of the nickel deposits was actually discovered by Thomas J. Flanagan, a blacksmith in the C.P.R. construction gang who noting a gossan showing on the right of way, dug into it and found copper and iron sulphides beneath the rusty covering. Subsequently the railway rock cut exposed the full width of the ore body. This deposit was later known as the Murray mine, after William and Thomas Murray who subsequently purchased the concession. Other discoveries followed in rapid succession, the last important deposit to be found (1916) being the Falconbridge mine now operated by the Falconbridge Nickel Mines Ltd.

The ore proved to be a complex mixture of three minerals, a nickel iron sulphide, pentlandite $(FeNi)_9 S_8$, a copper iron sulphide, chalcopyrite $(CuFeS_2)$, and an iron sulphide, pyrrhotite (Fe_7S_8), the amount of nickel varying from $1 \cdot 5$–$5 \cdot 0\%$. Many prospectors were attracted to the district, prominent among them being a certain S. J. Ritchie, who was chiefly interested in the material as a source of freight revenue, for a railway he had built in the district. He purchased a number of claims and in 1886 formed the Canadian Copper Co. to work the ore. This was the earliest Canadian corporate beginning of what is now the International Nickel Company of Canada. At that time, however, the promotors had no thoughts of nickel. Copper was their chief interest, and the presence of nickel in the Sudbury ores was not suspected. A contract was entered into by the Canadian Copper Company to sell one hundred thousand tons of the material to the Orford Copper Company (manager, R. M. Thompson) of New Jersey in the U.S.A. and mining operations were begun in 1886, the assay of ore shipped to Orford being copper $10 \cdot 2\%$, nickel $3 \cdot 5\%$, iron $32 \cdot 4\%$, sulphur $26 \cdot 0\%$ and silica $4 \cdot 2\%$.

The first shipments of ore did not respond to the usual methods for the extraction of copper, the product being a pale yellow metal not acceptable to the Orford Company's customers. Assay showed that this yellow metal contained nickel in addition to copper, and it became apparent that the smelting had repeated the experience of miners in Saxony operating on a similar material a century or so before. It was this experience that gave nickel its name. Frustrated in repeated attempts to produce copper from their ore, the Saxon miners named it 'Kupfer Nickel'—'Old Nick's copper'—and when Axel Cronstedt (1722–65) in 1751 succeeded in isolating the element which had caused the earlier difficulties, he employed the adjective as a noun and called it 'nickel'.

It was apparent that the success of the Sudbury venture was definitely dependent upon solving the metallurgy of the copper-nickel separation especially as a satisfactory means of extracting nickel would result in the ore being so much more valuable than if it contained copper alone. In those days the price of nickel ($1 per lb.) was about six times that of copper, the entire world production of nickel amounting to only about 1,000 tons, the bulk of which came from the mines of New Caledonia in the Pacific Ocean. On the other hand world consumption at that time was only about half the potential output from the Sudbury mines, so that developments of new markets was no less important than the discovery of a means of extraction.

A solution to the marketing problem was, however, soon discernible, for it was in 1889 that the classic paper of James Riley, a Glasgow metallurgist employed by Tennants (manufacturers of nickel on a small scale), described the improvements in the properties of steel armour plate that could be secured by the addition of nickel. The superiority of the armour plate so produced soon attracted the attention of the navies of all nations and demand for nickel began to outstrip production.

Although still lacking a satisfactory process for the separation of nickel and copper, the Orford Company were able to supply an oxide of nickel and iron suitable for incorporation in steel. The ore was first heap roasted and then smelted in a blast furnace for the production of a copper nickel matte. This was ground, roasted to eliminate sulphur, mixed with salt, again roasted to chloridize the copper which was then leached out with water. The insoluble residue from the leach was then dried and recycled, the final product being the twice leached insoluble residue consisting of the oxides of nickel, iron and some copper. As a considerable part of the nickel in the ore was also made soluble by the chloridizing roast and dissolved with the copper and lost to the process, the efficiency of the method was very low. A solution was eventually found in the use of nitre cake (sodium sulphate), a material then used as a reagent in the refining of copper. On tapping a charge of matte from the furnace to which sodium sulphate had been added it was observed that separation into two layers took place, an upper layer containing most of the copper and a lower layer of nickel sulphide containing but little copper. The upper layer was termed 'tops' and the lower 'bottoms', the process being known as the tops and bottoms process (Fig. 15). The process depends for its success on the fact that under the influence of the coke in the charge sodium sulphate is reduced to sodium sulphide, and that copper (and iron) sulphide is soluble in molten sodium sulphide to an almost unlimited extent whereas nickel sulphide is only slightly soluble. Further the specific gravity of the former is less than that of the nickel sulphide so that on solidification stratification occurs, the two layers being capable of separation by physical means such as blows from a sledge hammer. To increase the efficiency of the separation the treatment was repeated on the bottoms, the resulting second bottoms contain-

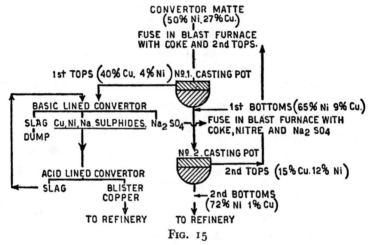

FIG. 15

ORFORD PROCESS FOR RECOVERY OF NICKEL AND COPPER

ing 70% nickel, 2% copper, 0·5% Fe. The 'tops' containing the copper were blown in a converter where the sodium sulphide was oxidized to sodium sulphate which was recovered and used again, the copper sulphide being blown to blister copper in the normal manner.

The bottoms were crushed, ground and then leached, first with water to remove the sodium sulphide and then with acid to remove iron.

The nickel sulphide was then given a chloridizing roast and leached to remove the copper, and a second roast with soda ash followed by a leach to remove remaining impurities. The resulting black oxide of nickel containing 77% nickel was then ready for despatch to the market or was mixed with coke and charged to a reverberatory where it was reduced to nickel of 98–99% purity.

The method was patented in 1890 and remained in use for sixty years being superseded in 1949.

Electrolytic Nickel. One of the first proposals to effect the electrolytic recovery of nickel was due to C. Hoepfner, who took out a number of patents in 1893–94. The process consisted in the treatment of powdered nickel copper matte with a solution of copper and calcium chlorides, whereby the copper and nickel go into solution as nickelous and cuprous chlorides. After removal of the copper by electrolysis, iron and any remaining copper were removed chemically, nickel then being deposited by a further electrolysis. Although much time and money were expended on the process in Canada it did not apparently prove a commercial success.

In 1899 W. Browne, of the Canadian Copper Co., developed an electrolytic process for the separation of nickel and copper. Nickel-copper matte produced in the blast furnace was roasted to eliminate sulphur and then melted down and cast into anodes containing about 55% Cu and 43% Ni. The electrolyte consisting of chlorides of copper and nickel was prepared by dissolution of part of the anode material in a solution containing chlorine and sodium chloride. Electrolysis took place in two sets of cells in the first of which copper was deposited on copper sheet cathodes, the nickel remaining in solution. When the concentration of the latter had reached a certain point, the solution was run off, any residual copper being precipitated with sulphuretted hydrogen, the liquor then being electrolysed at a higher potential in the second set of cells between carbon anodes and sheet-nickel cathodes. Nickel was deposited at the cathode with evolution of chlorine at the anode, the gas being collected and used for the dissolution of more copper and nickel anode material.

Electrolytic nickel deposition is complicated by the fact that nickel occupies a position relatively high in the electromotive series, requiring a high cathode potential for electrodeposition. This means that cations of elements rich in copper are more readily reduced at the cathode than nickel. Further, cations of elements such as iron and cobalt which occupy positions near nickel in the series are readily co-deposited. This difficulty was partly overcome by Browne, who used two sets of cells for the separate deposition of copper and nickel.

A Swedish metallurgist, N. V. Hybinette, overcame the problem by enclosing the cathodes in canvas-covered frames. The canvas was lightly woven so that a hydrostatic head was maintained in the frames to ensure a constant flow of electrolyte through the fabric at a velocity sufficient to carry away the copper and iron ions thus preventing them from migrating to the cathodes.

Actually, as the nickel ions were freed at the anodes they were not carried direct to the cathode but along with the copper and iron impurities were swept out of the cells as foul electrolyte, removal of iron and copper being effected in an external circuit, the purified solution being returned to the cathode canvas compartment. The nickel plated on the cathode was therefore supplied by the purified liquor constantly flowing into the cathode compartments.

The Hybinette process, operated at the Kristiansand refinery in Norway, was as follows. Nickel-copper matte containing 47–53% Ni, 28–34% Cu, 18–20% S, and 1% iron was roasted to eliminate sulphur and leached with 10% sulphuric acid solution to remove the copper, the insoluble residue being melted and cast into anodes containing 60–65% nickel. These nickel anodes were suspended in Hybinette bags, nickel being deposited on iron cathodes measuring 3 ft. × 3½ft. × ¾ in. The electrolyte contained 45 gm of Ni and about 1 gm of copper per litre was supplied to the cathode compartment, nickel being deposited. The electrolyte stripped of its nickel content then flowed through the diaphragm by virtue of its hydrostatic head to the anodes, taking up nickel, copper and iron and thence passed out of the cells and over waste anodes, which cemented out the copper and re-supplied nickel to the solution. (The amount of waste anodes obtained during electrolysis was about 30–40% of the original weight of the anode.) Nickel deposition took about ten days, the nickel sheets weighing 25–30 lb. The metal obtained assayed 99% nickel, 0·05% Cu and 0·5% Fe. The current density used was 8–10 amp. per sq. ft. at 3–4 volt per tank.

Any gold, silver, platinum, etc., present in the matte concentrated in the anode slime which was periodically collected for recovery of its precious metal content. The recovery of copper in the solution obtained from leaching the matte was effected by electrolysis.

The leaching procedure as applied in 1921 at the plant of the British American Nickel Corporation in Canada was somewhat different, the granulated matte being charged to tanks through which flowed the foul electrolyte from the cells, copper being cemented out, an equivalent amount of nickel being taken into solution.

The residual matte from this treatment containing about 45% Cu and 40% Ni was then roasted to remove the sulphur and leached with spent electrolyte from copper electrolysis. After clarifying, the liquor was pumped to the copper electrolysis unit. After the leaching the enriched nickel matte was mixed with coke and limestone melted down and cast into anodes assaying 70% Ni and 25% Cu.

Iron cathode plates were encased in Hybinette bags, the deposited nickel weighing from 20 to 30 lb. per sheet.

Mond Nickel. In 1890 a paper entitled 'Action of Carbon Monoxide on Nickel' and bearing the names of Ludwig Mond (1839–1909), Carl Langer (1860–1935) and Friedrich Quincke, was presented to the British Chemical Society. It described the formation of a gaseous compound of nickel and carbon monoxide by the reaction of the latter with finely divided nickel. The new compound corresponded to the formula $Ni(CO)_4$ and by heating the glass tubes through which the gaseous compound was passing, the carbonyl was decomposed with deposition and regeneration of nickel on the walls of the tube.

This paper was founded on an interesting fact. It appears that in the late 1880s Mond, who had established a chemical factory in the Midlands, was involved in a process for volatilizing ammonium chloride and, as the vapour attacked the majority of metals, difficulty was experienced in finding suitable materials. Nickel was experimented with for the valves, but it was found that these soon became coated with a black crust and ceased to function satisfactorily. The black deposit proved on examination to be carbon; it was found that the carbon dioxide used to sweep the ammonia gas out of the plant contained a small percentage of carbon monoxide, this latter gas being the cause of the valve attack. The ease with which nickel could be made to combine with carbon monoxide, and also the ease with which the nickel carbonyl gas could be decomposed with recovery of nickel, suggested to Mond that here was a possible method of refining nickel. Accordingly Mond applied for a patent covering 'The manufacture of Nickel'. A pilot plant was built at the works of Henry Wiggin and Company (Nickel Refiners) at Smethwick in 1892, using matte containing 18% Cu and 13% Ni bought from Sudbury, the process being transferred in 1900 to a much more convenient site at Clydach near Swansea. The part in the discoveries played by Quincke is not clear, but not long after the presentation of the paper he dropped out of the team leaving the long road from inception to industry to be traversed by Mond and Langer. The latter was an Austrian chemist and the carbonyl discovery was largely his. He later designed the equipment and for nearly thirty years superintended the intricate operations at Clydach.

In 1896 the process was brought to the notice of the Orford Company, which was then successfully separating nickel and copper by the tops and bottoms process and operating an electrolytic process for the production of pure nickel. None of the negotiations came to fruition. Another possibility explored by Mond was that of selling his patents to the Canadian Copper Company, for they held all the best mines in Canada, but had no refinery, their product being treated in New Jersey by the Thompson concern. These negotiations, however, likewise came to nothing.

Accordingly, in order to secure the raw material for the process, Mond purchased mining properties in the Sudbury area and erected a smelter for the production of matte for shipment to the refinery at Clydach where after the wet extraction of copper (as copper sulphate) his carbonyl process would be employed for the recovery of the nickel. To exploit this process, the Mond Nickel Co. Ltd., was formed in 1900.

In 1901 erection of the Canadian smelter was completed and copper-nickel matte was dispatched to Clydach. The matte arrived in the form of lumps and was ground to powder and roasted at 750° C. to remove the sulphur. The calcine was then leached with dilute sulphuric acid which dissolved out the copper, copper sulphate being obtained as a by-product by crystallization. The undissolved residue containing the nickel, after drying, was transferred to a series of tall cylindrical reduction towers, 6 ft. in diameter and 40 ft. high, each consisting essentially of a number of cylindrical sections or boxes, constructed so that the bottom enclosed a heating space. A central vertical shaft carried and rotated a series of scrapers in each box. Material entering at the top was worked across the boxes taking about 4 hours in all in its passage through the reducer. A temperature of 350–400° C. is maintained by the circulation through the heating spaces of hot flue gas resulting from the combustion of producer gas in a vertical brick furnace at the back. In its downward passage through the reducer the calcined oxide material is in contact with a stream of water gas—a mixture of hydrogen and carbon monoxide—which reduces it to finely divided nickel. At the low temperature employed nearly all the reduction is effected by the hydrogen ($NiO + H_2 = Ni + H_2O$) leaving an end gas rich in carbon monoxide suitable for use in the next stage. The reduced material is conveyed to another series of towers known as volatilizers similar in construction to the reducers—but with no heating arrangements, where it meets an upward current of carbon monoxide being converted to gaseous nickel carbonyl ($Ni + 4 CO = Ni(CO)_4$). To ensure complete volatilization of the nickel it is necessary to maintain a temperature of 50° C. and because the reaction is exothermic, the volatilizers are cooled. Eight volatilizers in all were used, the process taking several days. About 7–8% of the nickel, together with copper, iron and other impurities, remain in the residue. Decomposition of the gaseous nickel carbonyl is effected by passing the gas through a mass of continuously circulating nickel pellets contained in decomposition chambers maintained at a temperature of 180° C. At this temperature the carbonyl breaks down into nickel and carbon monoxide, the former being deposited on the nickel pellets as a thin coating, the carbon monoxide passing back to the volatilization plant for further use. In order to prevent the pellets from cementing together into a solid mass they are kept in movement by withdrawal at the bottom of the chamber and returning them to the top by means of a bucket elevator. Continual deposition of nickel produces a constant increase in size of the pellets and a more or less constant volume of material is maintained by providing an overflow pipe at the top. Pellets which pass through this pipe to a collecting box below constitute the production. Daily production per decomposer is some 1,500 lb.

the overall fuel consumption being 0·7 lb. per lb. of nickel. The pellets assay 99·9% nickel.

PRESENT-DAY NICKEL EXTRACTION

In 1949 the old 'Orford' tops and bottoms process at the Sudbury plant International Nickel was superseded by a new method which involved controlled cooling of the matte, followed by magnetic treatment and flotation. Many other innovations have also been incorporated in the flowsheet which can perhaps be best illustrated by a brief outline of present procedure.

The ore containing approximately 1% Ni and 1% Cu, together with small amounts of precious metals is crushed, ground and subjected to flotation. Three products are obtained: (1) a nickel-copper-iron concentrate, (2) a copper concentrate, (3) nickeliferous iron sulphide concentrate.

The concentrate containing 27% sulphur is fed to mechanical-type roasters (22 ft. diameter and 40 ft. high with ten hearths) the temperature being carefully controlled so that only half of the sulphur is expelled. The other half is needed to combine with the nickel and with the iron, otherwise some nickel will pass into the slag in the subsequent furnace operations. The calcine is discharged at a temperature of 700–800° C. and flows by gravity through pipes to the reverberatories. The stirring action of the roaster rabbles produces much dust, which is collected in a Cottrell dust-precipitation plant and incorporated in the reverberatory charge. The average charge to each roaster is about 250 tons of wet concentrate per day, including some 20% silica sand needed for fluxing in the next operation, advantage thus being taken to preheat the flux by the heat released during the roasting.

On subjecting the roasted ore to smelting a nickel-iron matte is obtained, the gangue and most of the iron reacting with the silica flux and being removed as slag.

$$FeO + SiO_2 = FeO\ SiO_2$$

The molten slag separates out on top of the heavier molten matte and is skimmed off at the end of the furnace directly into slag cars. It approximates 36% SiO_2, 37% FeO, and 7%

Al_2O_3. The matte, composed of 13% nickel and copper, together with cobalt, iron, sulphur and small amounts of precious metals, is tapped to a ladle transfer car.

Bessemer Conversion of Matte. The reverberatory matte, together with siliceous flux, is charged to a converter and blown to a 75% copper-nickel matte which is poured into ladles and transported to the adjoining department. The slag which contains 4·5% Cu Ni is returned for treatment to the reverberatories. The converters are of the conventional type, 13 ft. in diameter and 35 ft. long, lined with chrome magnesite brick. The matte cannot be blown to metal as in copper converting because the nickel would oxidize.

Separation of nickel and copper sulphides depends on a regulated slow-cooling of the matte from 926° C. to 520° C. in insulated refractory-lined steel moulds. As the temperature falls crystals of Cu_2S and Ni_3S_2 are formed substantially free from one another and capable of being mechanically separated by grinding. Precious metals of the platinum group concentrate during cooling in a metallic nickel-copper alloy. Solidification takes place at 926° C. when primary copper sulphide crystals form and grow as the temperature falls. As cooling proceeds a metallic-copper nickel alloy containing the precious metal crystallizes out, the formation of this alloy depending on careful adjustment of the sulphur content of the matte so that less is present than is required to combine with all the copper, nickel and other base metals present; the amount approximates to 75–90% of the amount required for complete sulphidization of the metals. When the matte cools to a lower temperature nickel sulphides form as separate crystals in addition to the copper sulphide and metallic alloy which continues to crystallize from the melt until complete solidification occurs. The cooled matte is crushed and ground and then subjected to a wet belt type magnetic separator, the metallic nickel alloy containing the precious metals being magnetic is separated from the non-magnetic copper and nickel sulphides. The magnetic alloy is transported to an electrolytic plant for recovery of nickel, copper and platinum metals. Separation of the copper and nickel sulphide constituents of the non-magnetic alloy is effected by selective flotation, the copper sulphide concentrate after filtration and drying being conveyed to the copper smelter, the nickel sulphide being dried and conveyed to the electrolytic department.

Electrolytic Refining. Up to 1958 the nickel sulphide flotation concentrate assaying 72% Ni, balance S, was mixed with coke and sintered, the nickel sulphide being converted to nickel oxide, the sulphur being reduced from 25 to 0·5%. Part of the sintered material was transported to the Clydach Nickel refinery in Wales for refining to metallic nickel (page 206), and part was mixed with coke and reduced to crude metal in a reverberatory furnace and cast into anodes on a casting wheel.

Refining of the anodes was carried out in mastic-lined concrete tanks, the electrolyte being a sulphate-chloride nickel solution containing 50–60 gm of nickel per litre. Cathodes were sheets of pure nickel plated out in special starting tanks set aside for the purpose. To prevent deposition of copper and iron impurities the cathodes are separated from the anodes by canvas diaphragms (page 204).

In 1958 it was announced that a new method had been developed for the electro-refining of nickel. The main feature is the direct electrolysis of nickel sulphide anodes which thus eliminates the sintering of the nickel sulphide and the subsequent reduction of the nickel oxide to crude metal as formerly operated. In addition to the substantial saving in the overall cost of refining, a further advantage is the recovery of contained sulphur which was lost during the sintering operation. This remains adhering to the uncorroded remnants of the anode as a voluminous cake. It assays 95% sulphur and is periodically collected and purified by fractional distillation.

In its essentials the electrolysis of the matte is similar to that of the older process, its main feature being the rigorous electrolytic purification system.

Electrolytic action at the anodes takes place at an anode efficiency of 95%, whilst the cathode efficiency of deposition is 99·5%. Thus a nickel deficiency in the anolyte is manifested by an equivalent increase in acid content which lowers the pH from 4·0 to 1·9. The correction of the nickel deficiency with restoration of pH value is achieved by carefully controlled addition of basic nickel hydroxide, followed by selective oxidation, hydrolysis and precipitation of iron, cobalt, arsenic and lead, removal of copper being effected by cementation on grain nickel. The purified electrolyte is then recycled to the cathodic compartments of the electro-refining cells. The cathode nickel (99·9% pure) is cut up into sizes required by the market and forms the main product of the refinery. Other forms such as ingots and shot are produced by melting and casting into the required form.

Treatment of Nickeliferous Iron Concentrate. The iron sulphide concentrate recovered from the flotation plant assaying 57% Fe, 37% S, 0·75% Ni, 0·05% Cu and 2·5% SiO_2 is treated for the production of a high grade iron ore and the recovery of its nickel content.

This involves five main operations.

1. Roasting for elimination of sulphur.
2. Selective reduction of nickel oxide.
3. Removal of nickel by leaching.
4. Recovery of nickel.
5. Pelletizing and indurating the magnetic leach residue.

The iron concentrate is suspension roasted in a fluid bed roaster, the sulphide sulphur of the iron mineral combining with oxygen in the air blast to form sulphur dioxide gas, whilst the iron and other metals form oxides. Part of the gas, laden with calcine, passes out of the top of the roaster to cyclones and electrostatic precipitators in which the calcine is collected. The gas contains about 12% sulphur dioxide, part being utilized for the manufacture of sulphuric acid and part being subjected to catalytic reduction with hydrocarbons for recovery of elemental sulphur. The majority of the calcine exits through the side of the reactor at a point opposite that of the feed and is collected in a separate system of cyclones, and fed by gravity to the kilns.

Reduction of the calcine takes place in rotary kilns 13 ft. in diameter and 185 ft. long, rotating at about 1 r.p.m. Reducing gas flows counter-current to the calcine, temperature and atmosphere control being maintained by partial combustion of oil or gas at the feed end of the kiln, the atmosphere being controlled to bring about essentially complete reduction of nickel (and copper) to their metallic states while reducing the bulk of the iron (Fe_2O_3) only to the magnetite state (Fe_3O_4). The material leaving the kiln is quenched in water and pumped to wet drum magnetic separators, the non-magnetics, mainly silicates, being discarded.

Leaching is carried out with ammoniacal ammonium-carbonate solution using five stages of counter-current decantation, about 95% of the extractable nickel being dissolved. The pregnant nickel solution is then aerated to hydrolyse and precipitate iron, filtered and then stripped of its ammonia by steam, the nickel precipitating as a basic carbonate. After drying the nickel carbonate powder is transferred to the electrolytic plant for recovery of metal.

The magnetite residue from the leaching plant is formed into pellets 1 in. in diameter, baked to facilitate handling and used as open-hearth charge ore by the steelmaking industry. The pellets contain about 68% iron.

PRESSURE LEACHING

Progress in nickel extraction has been marked during the past few years by the application of high-pressure leaching to the treatment of sulphidic ores containing nickel. The work was pioneered in 1947 by Professor F. A. Forward, and his associates at the University of British Columbia, and has now become an important branch of extractive metallurgy. The first plant utilizing the technique was erected in Canada at Fort Saskatchewan in 1953 for the treatment of nickel-copper-cobalt sulphide flotation concentrates from the Sherritt Gordon mines. The concentrate having the approximate composition Ni 12–16%, Cu 1–2%, Fe 30–40%, Co 0·5%, S 28–34%, is mixed with water and ammonia and subjected to a two-stage counter-current leach in large autoclaves (45 ft. long and 12 ft. diam.) under a pressure of 125 lb./sq. in. and a temperature of 940° C. to produce a solution containing nickel, copper and cobalt as ammines, together with free ammonia, ammonium sulphate, thiosulphate and polythionates. The iron present is precipitated as hydroxide and filtered off, the filtrate being boiled to reduce the ammonia content to a point where thiosulphate and polythionate are decomposed and react to precipitate the copper as sulphide which is filtered off. The copper-free solution containing nickel and cobalt is submitted to gaseous hydrogen reduction in further autoclaves at a pressure of 250 lb./sq. in. and a temperature of 175° C., nickel being precipitated.

$$Ni (NH_3)_2 SO_4 + H_2 = Ni + (NH_4)_2 SO_4$$

The precipitated nickel powder is filtered off, sulphuretted hydrogen precipitating the cobalt as sulphide from the filtered solution. The cobalt sulphide is then leached with sulphuric acid, any iron present being removed by oxidation, hydrolysis and filtration. Addition of ammonia converts cobaltous sulphate to the ammine, the cobalt then being recovered by treatment with hydrogen under pressure. An important by-product of the process is ammonium sulphate which is sold as fertilizer.

Recoveries are of the order of 90–95% for nickel, 80–92% for copper, 75% cobalt and sulphur 60–75%. About 30 tons of refined nickel a day are now being produced by the process at Fort Saskatchewan. The process has been established on a large scale in several other plants, the latest being at Moa Bay in Cuba. The ore which is a lateritic clay containing 1·3% nickel is currently being treated by the application of an acid pressure leach, annual production being scheduled at 25,000 tons of nickel per annum.

TIN

The middle of the 19th century saw Cornwall the centre of a flourishing tin industry, nearly half of world output coming from its numerous tin mines and smelters. Like the Welsh copper industry, however, the tin industry had by the end of the century dwindled almost to vanishing-point, the chief reason being competition from the more cheaply won tin from the rich alluvial deposits of the Malay Peninsula and other Far Eastern countries. It is to be noted that in their turn these rich deposits are year by year being worked out, leaner ores having to be exploited, with a consequent reduction in output. Thus it is possible that the high price for tin, £920 per ton (1962), may yet lead to a revival of Cornwall tin mining.

The only ore employed for the extraction of the metal is the stannic oxide mineral cassiterite (SnO_2) containing when pure 78·6% tin. In Cornwall and Bolivia the ore is found as primary veins or lodes in igneous or metamorphosed rock, invariably accompanied by such mineral impurities as iron pyrite, arseno-pyrite, galena, wolframite, etc. Cassiterite may also be found in river beds where it collects by natural elutriation in virtue of its high specific gravity (6·9), the lighter accompanying gangue material being carried away. Such deposits are known as stream tin and because of the action of the running water are in general free from contaminating minerals. Alluvial deposits are worked by gravel pump mining, panning, hydraulic sluicing, the richest deposits being worked by large electrically-driven dredgers weighing up to 6,000 tons, which may dig down to a depth of 140 ft. below water-level.

WORLD PRODUCTION (in tons) OF TIN IN CONCENTRATES

	1865	*1951*	*1960*
Bolivia	—	33,132	19,406
Cornwall	10,040	841	1,199
Malaya	6,000	57,167	51,979
Indonesia	5,280	30,986	22,607
Siam	2,000	9,502	12,080
China	500	7,500	24,000
Nigeria	—	8,529	7,675
Congo	—	13,669	10,109
Other countries	380	8,074	10,445
	24,200	169,400	159,500

Ore Dressing. Tin ores, whether alluvial or lode, are invariably too low grade to be smelted direct and must hence be concentrated. Alluvial deposits being comparatively pure are relatively easy to concentrate, a simple jigging operation sufficing to produce a 70% tin concentrate. In the case of lode tin owing to its contaminating impurities mineral dressing is much more extensive, gravity concentration yielding a 30% product containing Fe, As, Pb, W and Zn mainly as sulphides. Flotation, though not effective in the case of cassiterite, will effect separation of the sulphides. Prior to the advent of flotation (developed in the first decade of this century) roasting was employed to effect removal of sulphur and arsenic, the metals with which they were combined being converted into oxides. The roasted ore was then leached with acid to remove the oxides of Cu, Fe, Bi and Zn. Removal of Pb, Sb, Ag was effected by a chloridizing roast followed by acid leaching.

Wolfram undergoes no change and separation is more difficult. It may be extracted by fusing the ore with soda ash or sodium sulphate, when sodium tungsten is formed which being water-soluble may be removed by leaching. Wolfram may also be separated by magnetic means.

By a combination of these processes—gravity, flotation, roasting, lixiviation—the tin content of lode ore can usually be upgraded from 1% to between 55% and 70%.

Extraction of Tin. Reduction of tin from its oxide ore takes place on heating in the presence of carbon, according to the reaction.

$$SnO_2 + C = Sn + CO_2.$$

This apparently simple reduction is, however, complicated by three factors:

(1) The temperature required for reduction is very high (1200–1300° C.) as compared to the melting-point of the metal (231° C.); hence the oxides of other metals present are reduced at the same time and alloy with the tin.

(2) Cassiterite is amphoteric, i.e. it can act either as a base or an acid. If SiO_2 is present in the gangue some SnO_2 will combine with it forming a hard-to-fuse silicate, while if the ore contains lime a calcium stannate may be formed resulting in loss of metal in the slag. It is therefore necessary that the tin be freed as far as possible from impurities before reduction.

(3) Slags produced are too rich in tin to be discarded and have to be resmelted.

Smelting can be carried out either in blast furnaces or reverberatories. Provided that the ore is massive, rich and pure, blast furnaces using charcoal as fuel produce extremely pure tin and are admirable for the small worker, approximately two-thirds of the world's tin being produced in blast furnaces of small dimensions up to the beginning of this century. Compared, however, to the reverberatory, it suffers from many disadvantages amongst which may be listed

(a) Present-day tin ore is, as a result of mineral dressing methods, in a very fine state of subdivision and hence is not in a suitable physical state for charging to a blast furnace.

(b) There is much more loss by volatilization.

(c) When coke is used as the reducing agent the ash must be slagged off leading to the formation of much slag.

(d) Recovery of tin from slags is much better carried out in reverberatories.

(e) Reverberatories require no blast and permit the use of gaseous and liquid fuels. On the other hand the reverberatory siliceous hearth increases the slag output and the slags are richer in tin than those from the blast furnace.

Though the blast furnace was the original method and was in vogue for much longer than the reverberatory, being used in all tin-producing districts, today it is practically obsolete.

Reverberatory Smelting. The reverberatory furnace was first employed for the reduction of tin ore in Cornwall, from whence its use spread to all parts of the world. Pryce, in his description of the early reverberatory process, states the furnace differed little from that used in copper smelting. He gives the outside dimensions of the prevailing copper furnace as 18 ft. long, 12½ ft. broad, and 9½ ft. high, the interior dimensions being length 7 ft. 10 in., breadth 4 ft. 8 in., height 2 ft. and those of the fireplace 2 ft. 8 in. long by 2 ft. wide. He describes the process as follows: The charge which was introduced through a hole in the roof consisted of from 3 to 6 cwt. of tin ore mixed with 1/10th of its weight of anthracite culm. The fire was then raised to a great strength in which state it was left for 4–5 hours when the door was taken off and the charge well stirred, additional culm being added if required. After a further hour or so the metal was tapped to a basin made of clay. Part of the slag flowed out with the tin and that remaining in the furnace was raked out into a small pit made for the purpose. The tin in the basin was ladled out into moulds of about three-quarters of a hundredweight. The impure tin was then charged to a large (refining) furnace, a moderate fire put in hand, the molten tin making its exit through an open taphole into moulds being frequently stirred as it ran down leaving behind a dross. The furnace temperature was then increased and the dross and tin tapped out to a basin when the tin subsided, the dross rising to the top being taken off. Pryce puts the recovery as 12 cwt. of tin from 20 cwt. of ore.

Though this description refers to the process as carried out towards the end of the 18th century, in its essentials it could refer to present-day practice. Modern furnaces are on an average from 18–30 ft. long and 10 ft. wide, the hearth of the furnace being built over a hollow vault in order that it can be cooled by circulation of air beneath it, air holes being provided in the brickwork for this purpose. A cistern of water is sometimes provided below the hearth in which any tin that leaks through is caught. Owing to the excessive volatilization of tin an extensive system of dust and fume collection equipment is essential.

Operation. Present-day processes for the extraction of tin involve three main steps:

1. Reduction of ore with production of impure metal and a rich slag.
2. Treatment of slag for recovery of metal.
3. Refining of metal.

The ore charge is mixed with 15–20% weight of culm, and dross from refining operations. In about one hour a temperature of 1000° C. will have been attained which with repeated rabbling will have rendered the charge molten. Reduction readily takes place, making possible the tapping of metal continuously during the operation. Reduction is usually complete in 5–7 hours according to the amount of the charge and its quality.

Slag is directed to a separate pit containing water where it is granulated for convenience in handling and subsequent processing.

Treatment of Slag. The slag assays SiO_2 30–45%, FeO 15–25%, CaO 5–15%, SnO_2 5–25%, the tin being commonly present as silicate and as metallic prills. Treatment is effected usually by a precipitation method employing iron which at the temperature of operation decomposes the silicate forming ferrous silicate and liberating tin.

The charge consisting of slag, scrap iron, culm and dross is worked in a similar manner to the ore smelting but with a higher temperature. The products are a metal with a tin content of about 95% and a slag which is variable in composition, but may contain sufficient tin to justify resmelting. The metal obtained from this second fusion generally contains up to 20% iron and is known as hard-head. This is either returned with the ore charge or treated separately for the recovery of its tin content by decomposition with tin ore, the presence of SnO_2 initiating oxidizing conditions.

$$2\ SnFe + SnO_2 = 2Sn + 2FeO$$

Refining. The purity of the tin obtained in smelting depends on the purity of the ore smelted. Metal smelted from alluvial deposits generally requires very little further treatment yielding a metal containing 99·9%, but tin smelted from lode ores such as occur in Cornwall and Bolivia contain varying amounts of Cu, Sb, Pb, As, Bi, Fe, Ag and Ni, and hence necessitate refining. Present-day methods are refinements of the process already referred to on page 216, and consist of (*a*) liquation, in which pure tin is 'sweated' away from the high melting-point impurities, (*b*) boiling, the molten metal being subjected to agitation, the impurities being oxidized and forming a scum which is removed from the surface of the metal.

The impure tin is charged to the upper part of a small reverberatory furnace approximately 9 ft. long by 7 ft. wide whose hearth slopes down to the tap-hole, a temperature a little above the melting-point of tin being maintained. Pure tin liquates (melts) flowing down the hearth through the open tap-hole to the float leaving behind the more infusible impurities on the hearth. Fresh slabs are then charged and the process repeated until the float is full. The temperature in the furnace is then raised until the residue on the hearth is molten, when it is run off into another pot where it separates into two layers, an upper impure tin layer which is ladled out for further refining; the lower layer which assays 40–60% Fe, 15–30% Sn, 5–20% As, together with small quantities of copper, wolfram and sulphur is sent back for treatment to a slag furnace. The results where iron is the chief impurity are excellent but the separation of copper is not so effective because of the existence of a Cu/Sn eutectic of about 1% Cu.

Boiling involves the oxidation of the remaining impurities by injection of steam or air, the molten metal being violently agitated, impurities such as iron or zinc being oxidized and forming a dross. By the use of powdered sulphur the removal of any remaining copper can be effected and the separation of antimony and arsenic can likewise be achieved by addition of aluminium. Lead can be reduced by stirring in stannous chloride or injecting chlorine.

Boiling can also be induced by the introduction of logs of green wood, the heat of the molten tin decomposing the wood, gas and steam being given off, setting up violent agitation, thereby exposing the metal to the air. Tossing is also sometimes resorted to and consists of taking up a ladleful of the molten metal and pouring it back from a height.

The purity of refined tin may be assessed from the following assay:

Sn 99·93%. Fe 0·045%. Cu 0·018%. Sb 0·007%. Pb & S trace.

Electrolysis. Impurities may also be reduced to negligible proportions by electrolytic refining using a fluosilicate electrolyte. This method was first carried out by the American Smelting and Refining Co., Perth Amboy, New Jersey, U.S.A., who erected a plant in 1916 for the treatment of Bolivian tin concentrates.

The refinery consisted of seventy cells similar in size and construction to those used for copper refining. Electrolyte consisted of hydrofluosilicic acid of about 15% solution strength in which was dissolved 4% tin.

The crude tin was cast into anodes weighing 350 lb. each, cathode starting sheets being of pure tin ⅛ in. thick. After a week's operation the cathodes weighed 100 lb. and were removed from the tank, washed and melted down to pigs assaying 99·96% Sn. Three cathodes were produced per anode.

The electrochemical potential of lead in acid solution is so close to that of tin that both are deposited together; in order to prevent this happening sulphuric acid is added to the electrolyte to precipitate the lead as insoluble sulphate as soon as released from the anode. In order to obtain a coherent dense cathodic deposit it was necessary to use additive agents. Glue and cresylic acid were found to be effective.

The process operated successfully until 1923, when the plant was closed down, being unable to compete with cheap imported tin.

TINPLATE

Tin is used in large quantities for the production of alloys such as bronze, bearing metals and solder, but the greatest outlet is in the manufacture of tinplate, over one-half of total world production being so used. About two-thirds of this is used for food containers, the remainder in the manufacture of containers for oil, beer, paints, powders, tobacco, etc. The properties of tin which have led to the wide use of tinplate are its low melting-point which makes for ease of application, resistance to corrosion, non-toxicity, easy solderability and attractive appearance.

Early Developments. The idea of combining the strength of iron with the corrosion resistance of tin occurred early in history, the earliest tinplate being made by tinning wrought iron sheets laboriously pounded out by hand. Mechanical means of pounding the sheet into a sufficiently thin and flat form were eventually evolved by using a water-driven trip hammer called a 'helve', this development being followed by the application of rolling mills, water-driven, for making the sheet iron. The introduction of rolled iron sheets in the mid-18th century marked the real commencement of the technical development of the process of tinplate manufacture.

In 1829 a Welshman named Morgan patented the use of sulphuric acid for pickling, and thus eliminated the arduous manual descaling operation by rubbing with sandstone. This adoption of sulphuric acid in South Wales was probably due to the abundance of acid liquors from the neighbouring copper works.

An account of the manufacture of tinplate about this period was published in the Memoirs of the Literary and Philosophical Society of Manchester, the following being a brief abstract.

'The plates were rolled from bar iron of the finest quality known as tin iron made with charcoal instead of coke.* The sheets were then scaled by steeping in dilute muriatic acid and pickled in water in which bran had been steeped for some time and then in dilute sulphuric acid which gave the sheets a bright appearance. The pickled sheets were then placed in a grease pot and then in a tin pot for $1\frac{1}{2}$ hours or so, about 340 sheets being worked at a time. After removal from the tin pot they were redipped in a wash pot which contained about half a ton of metal, to ensure a more even coating of tin. The sheets were then brushed, given a final dip and then placed in a grease pot, the purpose of this being to take off any superfluous metal and

---

* The terms coke and charcoal are still applied to tinplate but are now used to indicate the thickness of the tin coating and not the quality of the basic metal.

thus even out the coating. The ends of the plates were then dipped in a 'list' pot to remove the excess tin from the list edge. After cleaning by hand the plates were ready for packing.'

The use of rollers to carry the sheets through the individual pots and also transfer between the pots was patented in 1878 by Morewood. This design which eliminated the arduous manual transfer is essentially similar in design to the Abercarn pot of the present day. Another important development about this period was the replacement of iron by steel-base sheets for tinplate. Actually the first steel tinplates were made about the year 1856 due to collaboration between Sir Henry Bessemer and Phillips and Smith of Llanelly. The business does not, however, appear to have progressed favourably. The true beginning of the manufacture of steel for tinplates can be attributed to Sir William Siemens at the works at Landore in South Wales which he started in 1870.

It is reported that the first tinplates produced in America were made in a copper works at Pittsburgh about 1858. In 1873 a few plates were made by a firm at Wellsville and another at Leechburg. Welsh prices were high and the industry in America could have been maintained, but in view of potential U.S.A. competition, the Welsh manufacturers reduced their prices with the result that firms closed down. They were not reopened for tinplate manufacture until 1890 when the Tariff Act commonly known as the McKinley Law, was passed. This law provided the imposition of a duty of 2·2 cents per lb. on imported tinplate. From that date the American industry grew rapidly and by 1912 production had outstripped that of Britain, U.S.A. production being 962,900 tons and U.K. 847,500 tons.

Production of the principal countries is as follows, in thousands of tons:

	1900	1939	1961
U.S.A.	303	2,617	5,009
U.K.	500	870	1,043
Germany	30	270	360
France	—	145	548
Japan	—	180	483
Others	67	418	1,573
	900	4,500	9,016

Modern Tinning. In the past tin coatings were obtained as already described by hot dipping, the uncoated steel plate in its finished size being passed through a bath of molten tin. Since the last war a second method has been developed which depends on the electrodeposition of tin on steel, the coating being continuously deposited on the surface of coils of strip which are then automatically sheared to plate size as they leave the machine. In addition to this continuity of operation a second advantage is that it permits the tin coating to be regulated to any desired thickness. These considerations led tinplate manufacturers in the U.S.A. to experiment on a large scale for some years before the war with the result that in 1944–45 a considerable proportion of America's tinplate production consisted of electro-tinplate which now accounts for well over 80%, the U.K. figure being about 70%.

Preparatory Treatment. The raw material for tinplate whether by electrolysis or hot-dip tinning is steel strip. Before it is tinned the strip is subjected to a number of preparatory processes in order to effect a satisfactory surface condition of the strip.

During hot rolling the steel strip becomes coated with oxide scale which has to be removed if a coherent tin coating is to be obtained. Formerly, this oxide scale was removed manually by sandstone but now the strip is passed through long tanks containing 6–12% acid, after which it is washed in hot water and dried by means of hot air. Electrolytic pickling is also employed, the strip being passed between two electrodes, between which an alternating current is passed, each side of the strip thus alternately becoming anodic and cathodic, so that both sides are pickled alike. By this method high speeds of pickling are achieved.

After pickling the strip passes through a cold reduction mill where it is reduced to the required gauge. Following cold reduction the strip will be hard and brittle due to work hardening; to soften the metal and render it ductile an annealing treatment is necessary. The modern electrically heated vertical annealing furnace functions at a temperature of 700° C. and at an operating speed of up to 300 ft. of strip per minute.

The strip produced by low temperature annealing is in the dead soft condition and if plastically deformed exhibits what is termed 'stretcher strain' which takes the form of kinking or fluting. The stretcher strain characteristic of this type of material is eliminated by imparting a small amount of cold reduction amounting to not more than 4%, the operation being known as temper rolling.

Electro-tinning. In this process the annealed steel in the form of a continuous strip made by welding long coils of strip together is made to pass through a solution of a suitable tin salt, electrolytic action causing tin to be deposited on the strip. Electrolysis takes place in long rubber-lined tanks 10–12 ft. deep which have at the top and bottom a series of rollers, the strip passing vertically up and down on the rollers in serpentine fashion travelling at a speed of some 300 ft. per minute. The anode consists of long bars of tin suspended vertically in the tanks, the cathode being the steel strip, the current leaving via the steel rollers.

Two types of bath are operated, one being designated the alkaline, the other the acid bath. A common alkaline bath is sodium stannate 120 gm/l and caustic soda 15 g/l, the bath being operated at a temperature of 70–80° C. Anodic replenishment of the tin content of the bath proceeds smoothly if the anode current density is kept sufficiently high and the free alkali sufficiently low to maintain a thin yellowish oxide film on the anodes. Under these conditions the tin enters solution as stannate; if, however, the anodic film breaks down, formation of stannite is favoured and the tin deposit impaired. The acid bath consists of stannous sulphate 50 g/l, sulphuric acid 60 g/l, cresolsulphonic acid 100 g/l and additive agents such as gelatine and B-naphthol. This bath has the advantage that it can be worked cold.

The amount of tin deposited is determined by the current density and the time of immersion. For normal working, the process is designed to give a depth of coating of 0·00003 in.

As it emerges from the electrolytic bath the strip appears both dull and lustreless and to achieve a bright silvery appearance it is passed over copper rollers through which current is introduced so as to heat the coating momentarily just above melting-point which has the effect of imparting a lustrous finish. The process is known as 'flow-brightening', and immediately after melting the strip is quenched in water to solidify the tin and minimize oxidation.

The flow-brightened strip is then sprayed with chromic acid to reduce any discoloration, steam dried and passed through an oiling machine where an emulsion of cotton seed oil is sprayed on to protect it during storage. Finally it is sheared into sheets and carried to a machine for stacking into piles.

HOT-DIP TINNING

In this process the pickled steel plates are fed into molten tin passing through a flux layer of zinc chloride which rests on the surface of the tin and which facilitates contact between the iron and tin, the molten tin 'wetting' the steel surface forming at the same time a continuous layer of $FeSn_2$. The plates pass through a series of guides and rolls immersed in the hot metal which serves to conduct the sheets through and out of the tin. The surface of tin on the exit side is covered with palm oil, which working in conjunction with the rollers serves to regulate the thickness and uniformity of the tin coating and reduce its temperature to a point below which oxidation would occur in the atmosphere. After emerging from the bath, the plates pass through a hot soda solution which removes the oil, and are then dried and polished by passing through felt-covered rollers.

Several different designs of pot are available for carrying out the hot-dip treatment. The Abercarn or Welsh pot is a 'double sweep' machine and may be considered as an advanced outcome of the older hand-operated tinning machine as patented by Morewood in 1878 in which the work was dipped first into a flux-covered pot, maintained at a relatively high temperature and then into an oil-covered pot working at a lower temperature.

The double sweep pot consists of two cast-iron pots connected together, the sheets passing through a flux layer into the molten tin and are then conducted by rollers and guides out of the bath and downwards into a second tin bath. Emerging from this second bath the sheets pass upwards through palm oil.

In the Thomas and Davies automatic combined feeding, pickling, tinning and finishing machine usually known as the Melingriffith unit the annealed steel sheets are conveyed automatically through the entire series of operations, being initially picked up individually by suction cups and fed to the pickling tank, the sheets being carried through on a large diameter wheel which conducts them on a half circular path, feed rollers then delivering them to the tinning pot. Heating of pots is effected by oil, gas or immersion-type electrical heaters.

The 'single sweep' machine often known as the Aetna or American-type machine follows the fundamental designs of that described by S. A. Davies in a U.S.A. patent issued in 1925. It consists of a steel box divided transversely by a partition extending to within a foot or so of the bottom, thus providing a passageway through which the plates pass from one section of the pot to the other, the sheets being only once immersed in the molten tin. The pickled sheets are fed to the pot at a uniform and rapid rate by means of magnetized rollers and then pass downwards through the flux into the tin by guides and rollers and then through the partition and upwards and out of the tin through the palm oil. This arrangement is capable of working at very high speeds.

Thickness and Weight of Coating. The coating weight is usually expressed as pounds of tin per basis box or as grammes of tin per square metre of tin plate. The basis box is the tinplate industry's unit of area and is equal to 31,260 sq. in. of tinplate, thus:

1 basis box is equivalent to 31,360 sq. in. of tinplate
      or   217·78 sq. ft. of tinplate
      or   56 sheets each 28 in. × 20 in.
      or   112 sheets each 20 in. × 14 in.

(Since each tinplate has two tinned faces the area of tinned surface in a basis box is 62,720 sq. in.)

It is generally accepted that 1 lb. per basis box is equivalent to 0·0000606 in. thickness of tin on each face of the tinplate equivalent to 0·036 oz./sq. ft. or 11·2 gm/sq. m.

The term coke generally represents the commonly used grades of tinplate with coating weight between 1 and 1½ lb. per basis box, charcoal plate averaging 2–3 lb. per basis box.

BIBLIOGRAPHY

COPPER

1. Trans. American Institution of Mining and Metallurgical Engineering. Vol. 106. *Copper Metallurgy,* 1933, New York.

2. *Handbook of Non-Ferrous Metallurgy*, D. M. Liddell (Editor-in-Chief). McGraw-Hill Book Co., New York.
3. PERCY, J., *Metallurgy*, Vol. I, 1861, John Murray, London.
4. *A History of Technology*, Vol. V, 1850–1900. Chapter 4. Oxford University Press.

ALUMINIUM

1. EDWARDS, J. D., FRARY, F. C. and JEFFRIES, Z., *The Aluminium Industry*, Vol. I, 1930. McGraw-Hill Book Co., New York.
2. RICHARDS, J. W., *Aluminium. Its History, Occurrence, Properties, Metallurgy and Application* (1896). H. C. Baird & Co., Philadelphia.
3. ANDERSON, R. J., *Metallurgy of Aluminium and Aluminium Alloys* (1925). H. C. Baird & Co., New York.
4. *Refining of the Non-Ferrous Metals* (1950). Institution of Mining and Metallurgy.
5. *Introduction to Aluminium and Its Alloys*. Aluminium Development Association, London.

ZINC

1. *Metallurgy of Lead and Zinc*. Vol. I, 121, 1936. American Institute of Mining and Metallurgical Engineers.
2. PERCY, J., *Metallurgy*, Vol. I, 1861. John Murray, London.
3. RALSTON, O. C., *Electro-deposition and Hydrometallurgy of Zinc*, 1921. McGraw-Hill Book Co., New York.

LEAD

1. *Metallurgy of Lead and Zinc*, 1936. A.I.M.E., New York.
2. PERCY, J., *Metallurgy of Lead*, 1870. John Murray, London.
3. HOFFMAN, H. O., *Metallurgy of Lead*, 1918.
4. GOWLAND, W., *Metallurgy of the Non-Ferrous Metals*, 1914. C. Griffin & Co. Ltd., London.

NICKEL

1. *Canadian Mining Journal*, 1937 and 1946.
2. *Story of Mond Nickel*. Privately printed and written for the Mond Nickel Co. Jubilee, 1951.
3. *Milling and Smelting the Sudbury Nickel Ores*. The International Nickel Co., New York, 1960.
4. Report of the Royal Ontario Nickel Commission, 1916.
5. FORWARD, F. A. and HALPERN, J., *Hydrometallurgical Processes at High Pressures*. I.M.M. Bulletin 603, February, 1957.

TIN

1. LOUIS, H., *Metallurgy of Tin*, 1911. McGraw-Hill Book Co., New York.

2. THIBAULT, P. J., *Metallurgy of Tin,* 1908. Sir Isaac Pitman & Sons Ltd., London.

3. MANTELL, C. L., *Tin, its Mining, Production, Technology and Application.* American Chemical Society Monograph No. 51.

4. HOARE, W. E. and HEDGES, E. S., *Tinplate,* 1944. Edward Arnold, London.

CHAPTER SIX

NEWCOMERS IN METALS

THE last few decades have seen metals enter new fields of expansion as the result of nuclear energy, guided missiles, jet engines and space technology. These usages involving extreme temperatures, radiation, high pressures and speed have led to the creation of new techniques in metal extraction and fabrication. Improvements in properties of the older metals have not served to satisfy the exacting demands imposed by these new environments, and metallurgists have had to turn to metals which were formerly considered too rare or too reactive. The increase in research into their extraction and refining has resulted in the development of such radically new techniques as ion exchange, solvent extraction, zone refining, electrode-arc melting, and high-pressure leaching. Many of these lesser known metals have long been known to metallurgists, but have hitherto found application in their own right to a very limited extent. Of the many factors which have delayed their exploitation, the relative scarcity is not necessarily the most important. For example, titanium is about 60 times as abundant as copper, and uranium is far more abundant than zinc, lead and tin. The difficulties have been rather in the problems both economic and metallurgical involved in processing the ores to metal.

To describe all the newcomers to the metallurgical scene and the techniques applied in their extraction and refining is beyond the scope of this volume. All that can be attempted is a representative sample of metals which best illustrate the progress and advances made in metallurgical technology. Recovery of uranium has led to the adoption of ion exchange and solvent extraction, titanium to the Kroll method for reduction of ore to metal and vacuum melting technique, germanium to zone refining. Cobalt has been selected, not so much as an example of the application of any new technique, but rather because of the diversity of metallurgical operations involved in reduction of the ore and refining the metal. The chief factors which have led to the emphasis on these metals have been the need to develop new sources of power provided by nuclear processes, the increasing stringent requirements of the aircraft industry for high-temperature metals, and a growing demand for metals for electronic applications.

URANIUM

Before the discovery that the rare natural-occurring uranium isotope U 235 would under nuclear fission liberate enormous amounts of energy, uranium had very limited usage, the most important being as a source of radium, small amounts of its compounds being also employed in ceramics to produce yellow glazes. In 1789 M. H. Klaproth (1743–1817), a German chemist and mineralogist investigating the pitch-blende deposits of the St. Joachimsthal mine in Bohemia (now Czechoslovakia), by reduction of the oxide with carbon isolated a black metallic powder which he assumed to be a new metal and named uranium in honour of Sir William Herschel's newly discovered plant Uranus.

About 50 years later E. M. Peligot in France, while examining Klaproth's black powder obtained by hydrolysis of the tetrachloride a total weight of chlorine which was greater than that of the original material assuming it to be metal. This led him to realize that the substance considered by Klaproth to be the element must in reality be some compound of the new element. In fact it was the dioxide (UO_2) derived from yellow uranium oxide. Peligot obtained the metal in 1841 by reduction of the tetrachloride with sodium and H. Moissan in 1896 obtained the metal in quantity by reduction of the oxide (U_3O_8) with charcoal in the electric furnace. For more than 100 years after its discovery uranium was used principally in the form of its compounds as a colouring agent for glass. Uranium salts were also used in the ceramic industry producing yellow and orange glazes.

In 1897 A. J. Becquerel discovered that uranium was radioactive, and this significant observation stimulated interest in this previously obscure element, and was followed in 1898 by the discovery of radium in pitchblende by the Curies. Actually radium is a disintegration product of uranium, the two always being found associated together. As a result of the increasing demand for radium in medicine, commercial sources of uranium ore were sought and were supplied for many years from the St. Joachimsthal mine.

Occurrence. About 1910 the uranium deposits of the Colorado plateau in U.S.A. began to receive attention. The uranium occurs as carnotite, a potassium uranium vanadate approximating $K_2O, 2UO_3, V_2O_5, 3H_2O$ containing 62–65% UO_2 when pure, and is found in a light-coloured sandstone. The deposits are of considerable extent (130,000 sq. miles) and for a time were the main source of radium. In 1913 the discovery of pitchblende occurred at the Katanga copper mine in the Congo, but because of the war and the urgent need for copper nothing was done about the discovery. In 1921 further development unearthed what proved to be the richest single source of uranium known in the world. Because of the high radium content (100–150 m.g. per ton) sufficient radium was soon being produced to supply all demands and at such low cost that it dominated the market, the monopoly continuing until the discovery of Canadian pitchblende deposits in 1930 at the Great Bear Lake in the Arctic Circle.

The following table gives some idea of uranium ore output in 1960 (excluding the communist countries):

Country	Production (tons) U_3O_8	Reserves (tons) U_3O_8
Canada	15,000	413,000
South Africa	6,200	330,000
United States	18,500	221,000
France	1,000	50,000
Australia	1,000	15,000
Congo	1,200	10,000
Total	42,900	1,039,000

In South Africa, uraninite, an oxide of uranium, was discovered in heavy mineral concentrates from the gold mines as long ago as 1923, but it was not until uranium became valuable as a source of energy that the potentialities of these deposits were realized. Most of the uranium at present produced in South Africa results as a by-product from the treatment of the tailing residues left after gold extraction by the cyanide plants. The ore grade ranges from 0·01 to 0·03 U_3O_8, extraction first commencing in 1945. Recognition of the geological similarities between the quartz conglomerates of the Rand and those of the Blind River district in Ontario, led to the discovery in 1953 in Canada of a vast uranium field, most of the mineralized rock containing over 0·1% U_3O_8.

EXTRACTION OF URANIUM

Before 1943 no metallurgical operation was conducted solely for the sake of obtaining uranium, recovery being effected for radium. The discovery of nuclear fission, however, completely changed the scene and occurrences which were uneconomic for radium became important as a source of the fissionable isotope U 235. Now deposits in which the uranium is present in as small amounts as $\frac{1}{4}$ lb. per ton of ore are being successfully exploited.

In the treatment of uranium ores in which radium is present in appreciable amounts, such as the heavy mineral pitchblende, the ore was formerly concentrated by gravity methods to yield a rich concentrate which was mixed with sodium sulphate and roasted in a reverberatory furnace to remove sulphur, arsenic and other volatile impurities. The roasting operation converts the uranium into sodium uranate. On washing with water the excess sodium sulphate and other soluble products were removed leaving the uranate in the residue with most of the gangue material. The residue was then treated with dilute sulphuric acid which dissolved the sodium uranate leaving a residue consisting largely of silica and radium sulphate. This was the starting-point for the extraction of radium. The acid solution containing the uranium was treated with an excess of sodium carbonate to precipitate the basic carbonates of iron, aluminium, nickel and cobalt, the uranium remaining in solution as sodium uranyl carbonate. On neutralizing with dilute sulphuric acid, and boiling, sodium diuranate separates as a heavy yellow precipitate which calcination transforms to U_3O_8. Carbon reduction yielded a crude metallic product.

Present-day Procedure. Uranium in ores may be solubilized after grinding by an alkaline or acid leach depending on the nature of the uranium minerals and the gangue present in the ore. When the gangue is siliceous little excess acid is consumed, but if the gangue is largely carbonate, acid consumption is prohibitive, and an alkaline solution is used. The bulk of commercial production is, however, obtained by processes using acid for leaching.

Alkaline Leach. When the gangue material is carbonate as is often the case with the Colorado Plateau carnotite ores, alkaline leach is the only feasible method. It is based on the fact that hexavalent uranium is soluble in excess sodium carbonate. $2U_3O_8 + O_2 + 18Na_2 CO_3 + 6H_2O = 6Na_4 UO_2 (CO_3)_3 + 12NaOH$. During the leaching operation it is necessary to control the hydroxide ion concentration in the solution because an accumulation of hydroxide precipitates insoluble uranium salts. This is secured by addition of sodium bicarbonate. Where applicable the alkaline method has some advantages in that it simplifies corrosion problems and requires relatively simple and inexpensive equipment; further, if vanadium is present separation of this element is comparatively simple. It has the disadvantage that it converts some of the silica in the ore to sodium silicate which renders filtration difficult, especially as the ore requires grinding to minus 200 mesh in order to render the ore soluble.

Acid Leaching. Sulphuric acid is most commonly used for this operation, generally in the presence of an oxidizing agent such as ferric iron which converts uranium to the hexavalent form, this compound being more soluble in water than when uranium has the valency 4 $(2U_3O_8 + O_2 + 6H_2SO_4 = 6UO_2SO_4 + 6H_2O)$. The ore is ground (Fig. 16) to minus 35 mesh and enough acid is used to bring the pH down to $1 \cdot 5$. The length of the leaching period is of the order of 10–20 hours, the acid consumption amounting to between 50 and 400 lb. acid per ton of ore. The insoluble material is separated by counter-current decantation and goes to the tailing dump, the clear uranium pregnant solution overflowing to storage tanks.

In the majority of cases the uranium is a very minor constituent in the solution, only 1 gm per litre or less. The bulk of the dissolved substances are iron, aluminium and manganese, which are present to an extent of two to three hundred times the uranium concentration. The problem that faced the

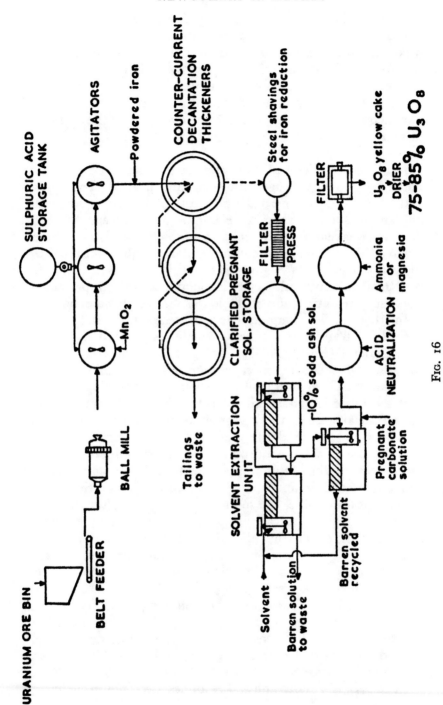

Fig. 16

Flowsheet of Uranium Recovery by Acid Leach and Solvent Extraction Process

industry in the pre-war period was to find a method of recovering the uranium present in such minute quantities and of producing it in a reasonably pure form capable of meeting the requirements of the atomic energy programmes. The difficulty was resolved by the development of two main processes, namely, the ion exchange method and solvent extraction. These are the most economic processes of recovering over 99% of the uranium in solution in a concentrated highly purified form.

Ion Exchange. The phenomenon of ion exchange was first noted by two British agricultural chemists, H. S. Thompson and J. T. Way, who in 1848 reported that on treating soil with ammonium sulphate most of the ammonium ions were absorbed and lime was released; the exchange of calcium and ammonium involved the exchange of equimolecular quantities of ions. In the years 1850–54, Way gave his results in lectures delivered to the Royal Agricultural Society in London and this represented the first systematic study of the ion exchange phenomenon.

In 1876 E. Lemberg, a German geologist, found it was possible to transfer the mineral leucite (K_2O, Al_2O_3, $4SiO_2$) into analcite (Na_2O, Al_2O_3, $4SiO_2$, $2H_2O$) by treating the mineral with a concentrated solution of sodium chloride, and that the transformation could be reversed by treating analcite with a solution of potassium chloride. The process emerged on an industrial scale at the beginning of the 20th century when R. Gans in 1906 used the method for softening water. The substances used for this purpose were natural minerals known as zeolites (sodium aluminium silicates). When hard water containing calcium and magnesium ions are passed through a column of the zeolite, the sodium ions of the zeolite pass into the water, whilst the calcium and magnesium ions are substituted in the solid. Thus the salts of calcium and magnesium responsible for hardness in the water are replaced by the corresponding sodium salts which are harmless, thereby rendering hard water soft. ($2Na X + CaCl_2$ (aq) $= Ca X_2 +$ $2NaCl$ (aq).) In time the water-softening capacity of the zeolite diminishes due to the removal of the sodium ions, but can be regenerated by passage of a sodium chloride solution, the sodium ions replacing the calcium and magnesium ions in the spent material.

In 1935 B. A. Adams and E. L. Holmes, two English chemists,

found that crushed phonograph records exhibit ion exchange properties. The study of this effect led to the observation that certain synthetic resins based on phenolic, sulphonic and amino structures could be prepared that would enable ion exchange to take place. This work was followed up in Germany and in the U.S.A. by G. F. D'Alelio and others and led eventually to the synthesis of resins from styrene, an organic material which has the property of forming chemical links with other styrene molecules to form long-chained material. By incorporating the correct amount of a cross-linking agent, small beads of resin, $0 \cdot 1$ in. in diameter, can be produced which are both mechanically strong and porous to water. Attached to this plastic matrix are active positively charged ionic amine groups imparting to the structure ion exchange properties. The resins can be visualized as consisting of an insoluble skeleton very similar to the body of a sponge, the skeleton being held together by chemical bonds (carrying a positive or negative electrical charge) which are balanced by mobile ions of opposite charge, the so-called 'counter' ions which are free to move in and out of the pores of the skeleton framework. When the sponge is immersed in aqueous solution, the counter ions migrate out of the pores, electric neutrality being maintained by ions in the surrounding solution diffusing into the pores of the resin. In other words this particular structure is capable of exchanging ions in the resin for similarly charged ions in the surrounding solution. Those resins which have negatively ionic groups attached to the matrix can exchange positive ions in a solution and are known as cation exchange resins. The resins with positive ionic groups can exchange negative ions and are known as anion exchange resins. The development of these synthetic resins made possible the economic application of ion exchange for the purpose of purifying and enrichment of metal-bearing solutions.

At an early stage in investigations involving methods of recovering uranium from leach liquor, the possibility of utilizing the ion exchange technique for isolating the uranium was considered. In South Africa during the period 1945–50, attempts were made to absorb uranyl cations selectively on a cation exchange resin. These attempts were unsuccessful since it was found that all the cations present in the solution were absorbed as readily as the uranium and hence no separation of this latter from the impurities could be achieved. Interest was revived in the ion exchange technique, however, when an important discovery was made in the United States in 1948. Investigations at the Batelle Memorial Institute, Columbus, Ohio, under the direction of H. E. Brosse, found that on passage of a solution of uranyl sulphate through a strong base anion exchange resin the uranium ions were selectively absorbed. The behaviour was simultaneously observed in the Dow Chemical laboratories in California.

This discovery immediately suggested the possibility of obtaining a selective adsorption of uranium from sulphate solution on an anion exchange resin, and experiments were conducted in which typical pregnant solutions obtained on the Rand were brought into contact with anion exchange resins. These experiments were particularly effective, and it was shown that, in spite of its low concentration, the uranium present could be selectively removed from the solution by adsorption on the resin. None of the other cations present as impurities in solution was adsorbed to any great extent, the ferric cation being the only other ion to exhibit a very slight adsorption due to the formation of a similar, but unstable, complex. It was also demonstrated that the uranium, once adsorbed on the resin, could be comparatively easily recovered by treating the uranium-loaded resin with a solution containing high concentrations of either nitrate or chloride ions. These ions could, by virtue of their high concentration, displace the uranium complex from the resin, producing a concentrated solution of uranyl ions containing only traces of other metallic impurities. Following this discovery a combined team of American and British workers went to South Africa in 1950 and set up the first resin exchange plant for the recovery of uranium at the Western Reef mine in the Transvaal.

In practice the pregnant solution is passed through beds of resin until the resin is loaded with uranium. After the loading cycle has been completed, the uranium is removed from the resins by the use of mixed solutions of ammonium nitrate and nitric acid or brine and hydrochloric acid. Uranium is then recovered from the solution by addition of ammonia or other basic compounds such as magnesia. The precipitate is then filtered and dried. Though there are no ion exchange columns in use at present for alkaline leach liquors, it is known that the method is technically feasible and investigations are proceeding.

A major drawback is that spent carbonate solution cannot be recovered for recycling.

The vessels used in the process consist of steel cylinders usually about 7 ft. in diameter and 12 ft. high, lined with rubber to resist acid attack. Only half the volume of the column is filled with resin as this doubles in volume during the process. Resins usually hold from 2–4 lb. of uranium oxide per cu. ft., from 400–800 lb. of product being recovered in each cycle. The columns are usually arranged in sets of three, with flow rates so adjusted that the resin in the first column is just saturated with uranium as the element begins to break through to the second column. The first column is now taken off stream and stripped with ammonium nitrate solution, whilst the liquor emerging from the second column goes through to the third. When uranium begins to show in the third, the second column is ready for stripping. Thus, while one column is being stripped, the other two are used to adsorb uranium, the process thus being semi-continuous. The uranium is recovered from the ammonium nitrate solution by precipitation with ammonia in the form of a bright yellow insoluble compound, ammonium uranate, which is then calcined to form uranium oxide containing 75–85% U_3O_8.

One of the chief advantages of the process is that the method concentrates the uranium in solution, bringing the uranium content from about 0·05% in the original leach liquor up to 0·5–2% in the final eluate. Further, the associated elements which would normally co-precipitate with uranium in a chemical precipitation are decreased or eliminated as these do not to any great extent precipitate from the eluate.

Ion exchange is also applied to other metal separations such as the rare earths and rare metals; uranium extraction was, however, the first large-scale application of the ion exchange method in metallurgy.

Solvent Extraction. This method of separation is based on differential solution, separation of the components of a liquid mixture being achieved by treatment with a solvent in which one or more of the desired components is preferentially soluble. The process, which is also known as liquid-liquid extraction, was first used on a commercial scale in 1907 by Edeleanu in America for the removal of aromatic hydrocarbons from kerosene. In this process kerosene is treated with liquid sulphur dioxide in a tower in which there is obstruction to flow such as perforated plates, ceramic or metallic rings. The kerosene by virtue of its lower specific gravity passes up the tower and meets the flow of the heavier sulphur dioxide coming down. The sulphur dioxide dissolves the aromatic components from the kerosene (raffinate) which is removed from the top of the tower, while the bulk of the solvent containing the aromatics is removed from the bottom. The kerosene, which now consists essentially of paraffin hydrocarbons, is a good quality material for lamps and stoves. The extract after separation from the sulphur dioxide, because of its high aromatic content and thus its good anti-knock properties, is suitable as a tractor fuel. The advantage of the process in this particular case is that the undesired aromatic compounds can be recovered unchanged and the refining agent can be separated and used again. The term extract is applied to that phase into which the solvent has been transformed and the term raffinate to that phase from which the solute has been extracted.

The process opened the field for the development of other solvent extraction processes such as the refining of lubricating oils and waxes, processing of pharmaceuticals such as the antibiotics, and more recently the separation and recovery of metals in solution.

Following on pioneer work done during the early 1950s at the Oak Ridge Laboratory of A.E.C. and at the uranium plant of the Kerr-McGee Oil Industries, New Mexico, the procedure has been found of value in uranium extraction; it improves on ion exchange in simplicity, speed and continuous operation.

The method is based upon the use of an immiscible solvent in which the component is preferentially dissolved. If an appropriate organic solvent is intimately mixed with an aqueous solution and allowed to settle, the substances present in the solution will distribute themselves between the organic and aqueous phases depending on the relative solubility in the two phases. The component with the greatest solubility in the organic solvent will tend to diffuse into this phase and thus a separation of the components originally present in the aqueous solution can be achieved. As the volume of solvent employed is considerably less than that of the aqueous solution enrichment takes place.

In its application to uranium the clarified pregnant solution is mixed with a solvent of the organo-phosphate type, dialkyl-phosphoric acid being commonly employed. Mixing takes place in an agitating tank, the uranium being transferred preferentially into the organic phase. The mixture of the organic and aqueous liquids flows to settling tanks when the lighter pregnant organic liquid collects on top of the aqueous solution. The process can be made continuous by a counter-current decantation system, (Fig. 16) employing a number of tanks, the solvent passing counter-current to the flow of pregnant acid solution. The dissolved uranium is then extracted or stripped out of the solvent by violent agitation with a 10% soda ash solution, the soda ash solution having a greater affinity for uranium than the organic solvent. The soda solution containing the uranium values is adjusted to a pH of 3 with sulphuric acid, addition of magnesia and ammonia precipitating yellow uranium oxide which after filtration and drying assays 75–80% U_3O_8.

The uranium oxide obtained by the above concentration processes contains small amounts of impurities. Peligot in 1842 discovered how to purify compounds of uranium by dissolving in nitric acid to obtain crystals of uranyl nitrate ($UO_2(NO_3)_2 6H_2O$) and then dissolving the nitrate in ether, the bulk of the impurities being insoluble. By evaporating the ether, Peligot obtained a considerable degree of purity. A modification of this method is still used, the etheral extract being treated with water, which results in the transfer of the nitrate to the aqueous phase from which it is precipitated by ammonia as pure ammonium diuranate.

Continuity of operation can be achieved by the use of a solvent extractive process employing methyl isobutyl ketone (hexone) or tributyl phosphate (T.B.P.) as the solvent. The uranyl nitrate solution is agitated with the solvent into which the nitrate is transferred. On settling, the lighter solvent collects on top and is drawn off, the nitrate then being transferred to the aqueous phase by agitation with water, from which it can be recovered in a highly purified state by evaporation. This is then dried and calcined to UO_3 and reduced to the dioxide (UO_2) by heating to $650°$ C. with hydrogen:

$$UO_3 + H_2 = UO_2 + H_2O$$

Fluidized Bed processes can be employed in the reduction,

the UO_3 being fed in a slurry to the fluosolid roaster, the reduced material overflowing to a bin.

Preparation of Metal. Early attempts consisted of reduction of uranium oxides with carbon, alkaline earth metals and aluminium and hydrogen, metal in various degrees of purity being produced. Impurities consisted either of reduction materials or substances the source of which was the container in which the reduction took place or atmospheric gases. Moissan in 1896 used 300 parts of U_3O_8 and 40 parts of sugar charcoal and reduced the mixture in a carbon container in an electric furnace. The crude product contained carbon and was purified by heating in a crucible lined with U_3O_8. R. Keeney in America in 1918 prepared an impure metal by reducing U_3O_8 with carbon in an electric furnace lined with magnesite walls, and obtained a product containing 88% U, the chief contaminant being uranium carbide.

Present-day procedure involves conversion of the dioxide (UO_2) to tetrafluoride by heating with hydrogen fluoride gas:

$$UO_2 + 4HF = UF_4 + 2H_2O$$

followed by reduction with metallic calcium or magnesium, a typical charge being 400 lb. UF_4 and 70 lb. Mg. Reduction takes place in a steel bomb 12–15 in. diameter by 40–50 in height, lined with a refractory commonly composed of dolomite.

Uranium can be cast and fabricated into desired shapes by conventional means such as rolling, extrusion and drawing. The metal is, however, very reactive and oxidizes at moderately high temperatures and as a consequence must be protected from air during fabrication. For use in atomic energy reactors the metal is cast into rods approximately 1 in. in diameter using vacuum melting technique in a high-frequency electric furnace. The rods are then heat-treated by annealing and quenching followed by tempering at 550° C. in a protective atmosphere of argon to relieve quenching stresses.

TITANIUM

In 1791 the Rev. W. Gregor (1762–1817), a noted mineralogist, whilst examining a black sand found in the Manaccan Valley, Cornwall, discovered a metallic substance which was attracted by a magnet. On analysis the material showed almost 50% of a white metallic oxide unknown to metallurgical chemists of that period. Gregor named the metal manachanite from the locality in which it was found.

In 1797 M. H. Klaproth identified manachanite with the substance which he had previously isolated in 1795 from a mineral, red schorl (rutile) found in Hungary. He named the new substance titanium in allusion to the Titans, the giants of mythology, and the name was well chosen for the metal proved to have several desirable and important features.

If the early employment of the metal had depended on its production in a pure state it would never have been used at all. In fact, the early utilization of titanium was chiefly in the form of its compounds, the oxide being used as a pigment, alloys of ferrotitanium being employed as scavengers in steelmaking. These two forms of the metal accounted for almost all the applications of titanium until comparatively recent times.

It is of interest to note that in 1859 R. Mushet (page 96) took out several patents for methods of utilizing ores of titanium and iron to obtain alloys of the two metals, thus foreshadowing the present use of ferrotitanium in the treatment of steel. In 1863 he formed a company, the Titanic Steel and Iron Company, to exploit his patents, but failed to convince the metallurgists of his day of the advantages in the use of titanium.

Titanium is the fourth most plentiful metallic element, ranking after aluminium, iron and magnesium. The principal titanium-bearing minerals are ilmenite, a titanium iron oxide (TiO_2FeO) and rutile, a natural oxide of titanium (TiO_2). Both minerals occur in igneous, metamorphic and sedimentary rocks, and in many cases are found concentrated in beach sands as in India, Australia, Brazil and Florida. Although still important sources of titanium, these beach deposits in recent years have tended to be replaced by massive rock deposits. Rutile is the main source of metallic titanium, ilmenite being used for the production of titanium pigments and ferrotitanium. Total world production of ilmenite is in the region of $1 \cdot 7$ million tons, and of rutile 100,000 tons, Australia being the main producer of the latter.

Throughout the 19th century repeated attempts were made to produce reasonably pure titanium from its compounds, but for the most part the metal was nearly always contaminated with nitrogen and carbon.

At elevated temperatures titanium has a great affinity for oxygen, nitrogen, carbon and hydrogen and absorbs limited quantities of these elements in solid solution which renders the metal brittle. This fact accounts for the principal difficulty in preparing high-purity metal and accounted for the early failures.

In 1887 L. F. Nilson and O. Peterson, two Swedish chemists, obtained metal by reduction of the tetrachloride with sodium in an airtight steel cylinder. The product contained $94 \cdot 7\%$ Ti, the chief impurity being oxygen. As this was present presumably combined with titanium, the material produced would contain about 79% titanium metal. Following up this work in 1910, M. A. Hunter at the General Electric Co. in the U.S.A., during the course of a search for filament material for electric lamp bulbs reduced titanium tetrachloride with metallic sodium in a steel bomb fitted with a lid held by braces and rendered gastight with a copper gasket to exclude air. The resulting metal had a purity of $99 \cdot 8\%$, melting in the region of 1800° C., which was far too low to be further considered as lamp filament material.

In the N–P method as modified by Hunter, the tetrachloride was made by chlorinating titanium carbide, the chloride then being refluxed in a current of nitrogen to remove excess chlorine. The fraction boiling at 136–137° C. was then shaken with sodium amalgam and redistilled. Five hundred gm of this plus 250 gm of sodium were then placed in the airtight steel bomb and brought to a low red heat, reduction of the chloride taking place. After leaching with dilute hydrochloric acid, titanium was recovered $99 \cdot 9\%$ pure in the form of metallic beads and powder.

In 1925 A. E. Van Arkel and J. H. de Boer, at the Philips Gloelampenfabrik in the Netherlands, developed the iodide dissociation process in which pure specimens of metal are obtained by sealing the material to be purified together with iodine in an evacuated chamber containing a heated filament of tungsten wire. On the application of heat the iodine reacts with the impure metal to form a volatile iodide which on contact with the heated filament thermally dissociates, depositing pure metal on the filament and liberating iodine which is available to produce more iodide. All danger of embrittlement with oxygen, nitrogen and carbon is thus avoided. The process is applicable on a small scale only and is not an industrial method. By the application of his process Van Arkel succeeded in preparing from sodium-reduced titanium quantities of metal 100% pure, on which physical and mechanical tests indicated that the metal had several desirable properties. Thus, although strong and corrosion resistant, the metal had only one-half the weight of steel, the combination of lightness with strength and high melting-point suggesting many possibilities in aircraft and other specialized uses.

Basing investigations on the method used by Hunter, Dr. W. J. Kroll (formerly in association with Siemens and Halske in Germany) of the U.S. Bureau of Mines succeeded in producing metal exceeding $99 \cdot 8\%$ purity and, following publication of his work in 1940, the development of large-scale production was actively pursued, notably by the U.S. Bureau of Mines, and by several companies in the U.S.A. and elsewhere. The E.I. du Pont de Nemours Co. of America developed the process on an industrial scale in 1948 and became the first producer of metallic titanium for general sale. Initial output was at the rate of 100 lb. a day at a price of $5 per lb. The first plant in this country was commissioned in 1949 by the Imperial Chemical Industries Ltd. at Birmingham and was designed to produce 12 tons a year in order to acquire experience in manuturing, melting and fabricating the metal. This was replaced by a 1,500 tons a year plant three years later. The original process was based on the Kroll method of reaction of titanium tetrachloride with magnesium, but was later replaced by a method based on reduction by sodium.

The original process developed by the U.S. Bureau of Mines employed reduction of the tetrachloride by magnesium metal $(TiCl_4 + 2Mg = Ti + 2MgCl_2)$. On completion of the reaction, liquid magnesium chloride formed in the reduction was separated by draining out the reaction chamber. The metal was chipped out of the reaction vessel, leached with dilute hydrochloric acid to remove magnesium and any remaining magnesium chloride, dried and ball milled to a fine powder, cold pressed at 50 t/sq. in. sintered under vacuum and then hot worked in protective steel sheaths (to exclude the atmosphere) to produce wrought metal.

Present-day procedure follows closely the above method. The raw material rutile (TiO_2) is first carburized by briquetting with coke, and heated, the carbide briquettes containing about 50% titanium then being chlorinated in a brick-lined furnace electrically heated to 800° C. Interaction between the titanium carbide and chlorine results in tetrachloride which distils over and is condensed in a cooling chamber:

$$TiO_2 + 2C + 2Cl_2 = TiCl_4 + 2CO$$

The tetrachloride is a yellowish liquid containing Fe 0·05–0·1%, Si 0·25–0·5%, V up to 2%, free chlorine 1·0–3·5% and non-volatile residues 1·7–6%. The impure chloride is purified by fractional distillation in the presence of copper or sulphuretted hydrogen.

The purified tetrachloride is reduced to metal by heating with molten magnesium in a steel reactor vessel located in a brick-lined furnace, provision being made for the introduction of an inert gas such as helium or argon to eliminate the reactivity of titanium with the atmospheric gases, oxygen and nitrogen. Reaction proceeds according to the following:

$$TiCl_4 + 2Mg = Ti + 2MgCl_2$$

By electrolysis of the by-product magnesium chloride, magnesium and chlorine are obtained and recycled for further use. Titanium in the form of metal sponge accumulates in the reactor, magnesium chloride being tapped off from time to time.

Magnesium is the usual reductant, but sodium, first used by the Degussa concern in Germany during World War Two, is employed at the I.C.I. Metal Division plant at Witton, Birmingham. Advantages claimed are (1) Sodium (m.p. 98° C.) can be introduced in liquid form into the reactor during the reaction process. It is preferential from both physical and chemical viewpoints to add the reducing agent at the same rate as it is consumed. Addition of magnesium (m.p. 650° C.) in liquid form is more difficult. (2) Owing to its lower melting and boiling-points, sodium is easier than magnesium to purify by distillation. (3) As sodium chloride is not so hydroscopic as magnesium chloride subsequent handling is simplified. (4) Sodium is cheaper than magnesium.

The metal obtained in the magnesium reduction process has the appearance of porous pebbles, whilst that from sodium reduction has the form of dull grey granules. In either case the metal has to be consolidated into massive form.

The high melting-point (1670° C.) and the high reactive nature of the metal combine to make the melting and casting of titanium a difficult problem. The molten metal reacts with all the usual refractory materials normally used to make or line a furnace; moreover, molten titanium absorbs gases, notably oxygen and hydrogen, which affect its physical properties and lessen its usefulness as a structural material.

Arc Melting. In 1905 Dr. W. von Bolton of Siemens, in Berlin, evolved a method of melting tantalum (melting-point

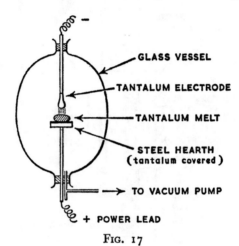

GLASS VESSEL

TANTALUM ELECTRODE

TANTALUM MELT

STEEL HEARTH
(tantalum covered)

TO VACUUM PUMP

+ POWER LEAD

Fig. 17

Von Bolton Electric Arc Melting Furnace

3000° C.) in which an electric arc was used as the source of heat, an arc being struck between an electrode and the metal. The apparatus used (Fig. 17) consisted essentially of an evacuated glass vessel containing an electrode composed of tantalum, and a steel hearth. The compacted metal was placed on the hearth, a vacuum applied to the vessel and direct current applied to the electrode, the arc being directed on to the metal contained in the hearth. The metal consolidated into an ingot which on examination proved to be soft and ductile. The discovery of the principle that metals with high melting-point and gas sensitivity could be successfully arc-melted constituted a great advance.

W. J. Kroll in 1932 saw the arc melting operation at the Siemens and Halske plant in Berlin, was greatly impressed with the technique and in 1940 applied a modified form of the method to the consolidation of sponge titanium into massive metal. He employed a tungsten tip for his electrode and replaced the steel hearth used by Bolton by a water-cooled copper crucible thereby taking advantage of the high thermal conductivity of copper, the molten titanium freezing on the crucible wall forming a container for the metal.

It will be remarked that the electrode used by Kroll was of a

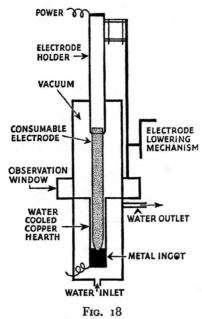

FIG. 18

CONSUMABLE ELECTRODE VACUUM ARC FURNACE

permanent nature being constructed of tungsten, the metal to be melted being contained in a crucible. This is known as a non-consumable type of electrode. It was later found that contamination of the titanium was liable to be caused by tungsten from the electrode and this process was replaced in 1954 by a consumable electrode. In this type the electrode was made by welding together compacted sponge titanium to form round or square electrodes. This electrode is then suspended from the top of the furnace (Fig. 18), while the water-cooled copper crucible is clamped on to the bottom. The furnace is evacuated and an arc struck and the whole electrode, slowly lowered by an electric motor, melts to form an ingot, the electrode metal

246

melting and passing to the ingot mould, in the form of a fine spray. Melting proceeds up to a rate of 100 lb./min. in a 20 in. diameter furnace, ingots weighing up to 20 tons being produced. To ensure complete homogeneity the process is repeated, the ingot being raised to the top of the furnace, the crucible charged and the ingot remelted. In addition to titanium the method is now being used for zirconium, molybdenum, steel and nickel-base alloys.

Hot working of the metal is complicated by the fact that at elevated temperatures the metal tends to take into solution hydrogen and oxygen from the air, which causes embrittlement. Further, the metal has a low heat capacity which makes it lose heat rapidly and makes hot working difficult. Due to high friction between the metal and steel it is difficult to roll, reduction for a given load being less than for stainless steel. However, provided that due allowances are made for these characteristics, conventional metal-working techniques can be employed to forge the metal into flat, slab or round billets preparatory to rolling.

Electrolysis. The possibility of producing titanium by alternate methods avoiding the costly thermal treatments and batch nature of the process has prompted enquiry into electrolytic methods. Research, has, however, shown a number of difficulties which may be summarized as follows:

(1) Titanium cannot be desposited from aqueous solutions, hence fused salt electrolytes have to be employed.

(2) Choice of salts is restricted to those free from oxygen and with a cation which is not reduced by titanium. Further, there is only a limited number of titanium compounds which are soluble enough in fused salts and cheap enough for practical application.

(3) Titanium cannot like aluminium and magnesium be deposited in the molten state because of its high melting-point.

(4) Owing to the sensitivity of titanium to impurities the electrolysis has to be carried out in a very high purity electrolyte and in an inert atmosphere. This tends to complicate the design of any apparatus for continuous operation.

Some success has, however, apparently been achieved, for a number of projects have been developed up to the pilot plant stage.

In 1951 G. D. Cordner and H. W. Worner in Melbourne employed titanium trichloride prepared by hydrogen reduction of the tetrachloride ($TiCl_4 + \frac{1}{2}H_2 = TiCl_3 + HCl$). The trichloride was dissolved in a salt bath composed of lithium and sodium chlorides and electrolysed at 550° C. under a hydrogen atmosphere using tungsten electrodes. Globules of metallic titanium were produced using a current density of 6 A/cm² with a current efficiency of 30–60%.

Following this up, the Titan Co. Inc. of America used a two-stage cell containing one anode and two cathodes and an electrolyte composed of 7 parts strontium and 27 parts sodium chlorides at a temperature of 560° C. In the first stage titanium tetrachloride was reduced to a lower chloride at one of the cathodes, the chloride then being reduced to titanium metal at the second cathode under an inert gas cover. A semi-adherent cathodic deposit of metal assaying 99·5% titanium was produced at a current efficiency of 90%, with a relative low expenditure of power—5·5 kWh per lb. of Ti produced. Difficulty was, however, experienced in obtaining coherent deposits and no further developments of this process have been reported.

A. Brenner and S. Senderoff in America in 1952 employed potassium fluotitanate (K_2TiF_4) as the raw material, a bath composed of 5 parts each of sodium and potassium fluorides operating at 850° C. being used. It is reported that the Horizon Corporation of Stamford, Conn., who operate an electrolytic process for production of zirconium have developed the process to the pilot mill stage.

Alloys. Pure titanium is characterized by low tensile strength (14 tons/sq. in.) and high ductility (55% in 2 in.). Industry demands a metal with a wide range of mechanical properties and these requirements are being met by incorporation of alloying elements. Manganese, aluminium, molybdenum, vanadium, tin and chromium in varying amounts and combination have been found to lend versatility to the mechanical properties of ductility.

Up to a temperature of 885° C. titanium has its atoms arranged in the hexagonal close-packed structure (alpha phase) and is relatively soft, tough and ductile. Above 885° C. the atomic structure changes to the body-centred cubic structure, metal in this beta phase being hard, strong and less ductile. Based on these allotropic modifications the alloys of titanium can be divided into two categories, alpha alloys and alpha-plus-beta alloys. Metallic elements aluminium, zirconium and tin, together with oxygen and nitrogen all have appreciable solubility in the alpha phase, alpha alloys being accordingly based on these elements.

Most commercial alloys are alpha-plus-beta, the presence of the beta phase facilitating forging; this class of alloy has also the advantage of responding to heat treatment. Incorporation of 2% Mn, 2% Al and annealing give rise to an alloy possessing the tensile strength of mild steel with the added advantage of possessing only half the weight. Doubling the alloy additions increases the tensile strength to 70–80 tons/sq. in.

A 6% Al, 4% Va responds to heat treatment and may be used for highly stressed components. It has the advantage of being weldable.

A 5% Fe, 0·1% N_2, 0·4% C may be hot forged and quenched, resulting in an alloy possessing a tensile strength of 100 tons/sq. in. but with only 1·5 elongation.

Usage. The properties of titanium and its alloys make it particularly suited to the needs of the aircraft industry for a light alloy which could meet the strains imposed by ultra-high-speed flight. It has an exceptionally high strength/weight ratio which is retained at temperatures higher than are considered safe for aluminium alloys, and satisfactory creep properties can be attained at temperatures as high as 500° C. The fatigue ratio of titanium and titanium alloys is appreciably higher than that of most non-ferrous metals. Commercially pure titanium has a low density and will withstand moderate temperatures for long periods or high temperatures for short periods, while its resistance to attack by sea water makes it suitable for applications where immunity to corrosion is important.

For the chemical and electro-chemical engineering industries titanium has very useful properties in its outstanding resistance to corrosion and erosion in a wide range of aggressive media, such as nitric acid, aqua regia, chlorine, metallic chlorides and organic acids. Anodes made of titanium with an extremely thin coating of platinum can carry very heavy anodic currents, offering the advantages of platinum at much lower cost.

Titanium-platinum anodes are now in use in electrolytic cells for chemical manufacture and for the cathodic protection of chemical plant and marine structures. The metal is completely inert in hot or cold chromic acid, nickel plating solutions and most other liquors used in metal finishing.

GERMANIUM

While visiting a German zinc smelter in 1945, the author was shown a small piece of greyish white metal which had been isolated from zinc ore many years previously, and was informed that this was germanium. The metal had no usage and neither the author nor his German colleagues realized then that this little known laboratory curiosity was in a very short space of time to open up a new era in the electronic industry. Research showed that the metal had semi-conducting properties, its electrical conductivity being intermediate between that of metals and insulators. When used in conjunction with suitable electrical contacts some of the ions are free to move giving rise to a current in a particular direction, current flowing much more readily in one direction than the other. In effect the current becomes uni-directional, alternating current being rectified to direct. This is the principle of the crystal diode rectifier. The Bell Telephone Co. of the U.S.A. in 1948 showed that minute slices of suitably prepared germanium can be arranged in such a manner that they are capable, under the influence of an electric field, of amplifying currents much in the same way as in an ordinary radio thermionic valve. As current is transformed across a resistor this device became known as a transistor. Its advantage over the thermionic valve is that having no filament to heat, no high tension supplies are required. Moreover, its small size and robust construction have many obvious technical and economic advantages. Over 200 million semi-conductor diodes and triodes are now being produced every year and it appears certain that their scope will be considerably extended as research succeeds in bringing down their cost.

Germanium is generally found as a sulphide associated with other minerals in particular those of zinc, and is recovered as a by-product during the process of recovering the primary metal. The first major recovery of the metal was made in 1952 from the zinc ores of the Tri-State (Missouri, Oklahoma and Kansas) deposits in the U.S.A. The zinc ores of Katanga in the Congo have also in recent years yielded significant amounts of germanium. On sintering the zinc ores, cadmium and germanium are volatilized and collect in baghouses or electrostatic precipitators, this material forming the raw material for production of the metal. The British source of the metal is the dust found in flues of gas-producer plants. Many British coals contain germanium of the order of 10–12 parts per million. During the process of using the coal to make producer gas, germanium in the form of oxide or sulphide passes out with the gas and on burning the gas the dust accumulates in the flues containing up to 1% Ge. A preliminary concentration is given by smelting in a small reverberatory furnace with coal dust as a reducer, soda ash and lime, slagging off the alumina and silica, copper oxide being present to collect the germanium in a regulus. This regulus containing about 4% germanium is granulated by pouring into water and then subjected to treatment with chlorine in the presence of ferric chloride. Distillation of the chloride solution in the presence of sulphuric acid results in germanium tetrachloride containing up to 15% arsenic trichloride. This forms the raw material in the production of the metal. Production in Britain was initiated in 1950 by Johnson Matthey & Co.

Earlier History. In 1886 C. Winkler, assaying a sample of a silver sulphide mineral known as argyrodite from Freiberg in Germany, obtained 74·7% Ag, 17·1% S, 0·66% FeO, 0·22% ZnO and 0·31% Hg, the result being 7% too low. He attributed the difference to an unknown element which precipitated as sulphide in the H_2S group. Winkler finally isolated the element and named it Germanium after his native country.

Winkler isolated germanium from argyrodite by heating with soda ash to render soluble, extracting with water and treating the solution with sulphuric acid, the precipitated sulphur, arsenic and antimony being filtered off. The solution was treated with hydrochloric acid as long as a precipitate formed and then saturated with H_2S and filtered, the sulphide precipitate being washed with alcohol containing H_2S. The sulphide was then roasted at a low temperature, warmed with nitric acid and the oxide so obtained ignited to metal by heating in a current of hydrogen. The result was a greyish white metal melting at 958° C.

In 1917 G. H. Buchanan devised a process for separating the element obtained from zinc ores based on the volatility of the tetrachloride ($GeCl_4$). The crude material was heated with hydrochloric acid in a distillation flask, and the distillate treated with H_2S. The germanium compound so obtained was contaminated with arsenic, which is an invariable constituent of all germaniferous ores. In the above procedure the arsenic volatilizes as trichloride and L. M. Dennis in America prevented this from taking place by passing a current of chlorine through the apparatus during the distillation. This maintains the arsenic in the quinquevalent condition which does not volatilize. He further found that if the flask be fitted with a fractionating column, the $AsCl_3$ boiling at 122° C. can be separated from the $GeCl_4$ which boils at 85° C.

For the electronic industry a very high degree of purity is essential, in particular the arsenic content must be extremely low, of the order of 1 part per million. Fractional distillation, though effecting a separation, does not completely free arsenic, since arsenic trichloride is appreciably volatile in the vapour distilling from a solution of germanium tetrachloride in hydrochloric acid. Bubbling chlorine through the solution tends to keep most of the arsenic in the non-volatile quinquevalent form but a back reaction invariably occurs with the formation of arsenic trichloride and chlorine. The only satisfactory method of eliminating arsenic is to reduce it to the metallic state, but care has to be taken to avoid at the same time reducing the germanium tetrachloride to the dichloride which is not volatile. The modern approach is to use metallic copper with which arsenic trichloride reacts to form cuprous chloride and copper arsenide, whereas germanium tetrachloride is unaffected. Refluxing for some time over copper turnings eliminates the arsenic.

Hydrolysis of the separated tetrachloride with ice water results in the precipitation of the dioxide which is washed and dried. The reduction to metal is achieved by heating in a current of hydrogen:

$$GeO + H_2 = Ge + H_2O$$

Reduction takes place in two stages, the dioxide being first heated to below 700° C. above which temperature, the monoxide which is first formed is volatile. When water ceases to be evolved the temperature is slowly raised to 1000° C. at which temperature the powder consolidates into metal.

Zone Refining. The metal still contained traces of such impurities as arsenic, antimony, indium, sulphur and iron, and further purification is necessary in order to achieve the very high level of purity required for electronic needs. This has been achieved by 'zone refining' introduced in 1952 by W. G. Pfann of the Bell Telephone Co. of U.S.A. The method is based on the fact that in many metals segregation of the impurities between a molten and solid part of the material can be induced under

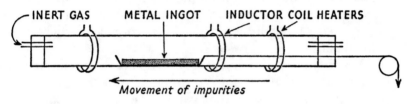

INERT GAS   METAL INGOT   INDUCTOR COIL HEATERS

Movement of impurities

FIG. 19.

ILLUSTRATING PRINCIPLE OF ZONE REFINING

the influence of heat, the impurities being more soluble in liquid than in solid metal. Thus, if an ingot of metal is fused for a narrow portion of its length and this zone of fusion is moved along the ingot (Fig. 19), pure metal will concentrate to the cooling portion whilst the impurities will be carried forward in the molten portion towards one end of the bar in the direction of zone movement. By a repetition of this process it is possible to obtain a metal in which most of the impurity is concentrated at one end, the purity increasing with distance towards the other end.

In its application an ingot of the metal contained in a crucible of pure graphite or pure silica is placed in a long quartz tube through which is passed an inert gas such as argon. The tube is surrounded at intervals by three or four electrical induction coils which are arranged to travel together slowly from one end of the ingot to the other, so causing three or four molten zones about an inch wide to move along the ingot in one operation.

The impurities present tend to remain in the liquid metal rather than move across the liquid/solid interface at the trailing end of the zone and this results in their being swept to that end of the ingot towards which the coils are moving. The zone containing the impurities is then discarded by cutting off the end at a suitable point. The effectiveness of the method can be shown by the fact that arsenic originally present to the extent of 5–10 parts per million can be entirely eliminated, while iron, sulphur and antimony can be reduced to below 1 p.p.m.

An alternative method suitable for metals which are liable to be contaminated by the material of the crucible is the use of 'vertical zone refining' developed by P. H. Keck in 1953. In this method, the ingot is gripped at both ends in a vertical position, the surrounding container being evacuated.

A refinement has been the employment of the electron beam as a source of heat, a stream of electrons obtained from a heated filament being focused by means of electro-magnetic lenses on to the metal.

Although the first important application of zone refining was for purifying germanium for electronic use, the method can be applied to any substance that can be melted, and exhibits a difference in impurity concentration between liquid and solid. Niobium, zirconium and titanium can all be obtained in a high state of purity by the method. Organic substances such as naphthalene and fatty alcohols are also being treated by the technique.

COBALT

Although the use of cobalt compounds as a colourant in the manufacture of glass and ceramics dates from ancient times, its industrial applications are comparatively recent. The discovery and practical application of the most important specific properties of metallic cobalt go back only 40 years. For instance, its incorporation in high-speed steels and permanent-magnet alloys became current only after World War One. Since then new properties have been brought to light leading to numerous applications in the most varied fields.

The earliest recorded use of the word cobalt appeared in Agricola's *Bermannus* in 1530 and was used to designate a substance found in the silver mines of the Erzgebirge region in

Saxony. On the other hand at the same period in another region of Germany—the Harz mountains—certain copper ores when roasted not only failed to yield metal, but emitted poisonous fumes from the associated arsenic and were referred to by the miners as *Kobold*, the German term for goblins and gnomes, on account of the mischievous effects on health. These ores were roasted to eliminate the arsenic and sulphur and yielded a mixture of crude cobalt oxide called zaffre which on fusion with potassium carbonate and sand produced a blue glass-like material known as smalt. This was crushed and ground to a fine powder and used by glassmakers to colour their wares. Smalt was used in glass works until quite recently when purer forms of cobalt compounds have supplanted it.

G. Brandt, a Swedish chemist, was the first to show that the blue colour of smalt was due to cobalt which he isolated in 1735 by reduction of the ore. Brandt also described the properties of the metal and it was he who discovered its magnetic behaviour. The elemental character of cobalt was subsequently established in 1780 by Bergman.

Occurrence. From the 16th until the middle of the 19th century, the world's supply of cobalt ores came from Norway, Sweden, Saxony and Hungary. In 1864 oxidized cobalt ores were discovered in New Caledonia and became the world's main source of cobalt until about 1904, when the silver cobalt ores of Ontario, Canada, led to the virtual closing down of production in New Caledonia. The rich ores found in the town of Cobalt in Northern Ontario formed the main source of supply until the discovery of the Katanga deposits. From thence until World War Two cobalt recovery in the Ontario area received only spasmodic attention, silver being mainly sought. In recent years, however, Canadian production has shown a steady increase, cobalt being derived as a by-product from the nickel industry at Sudbury and from pressure-leaching operations of nickel-copper-cobalt sulphide ore conducted by the Sherritt Gordon Mines, Alberta.

In 1914 cobalt-containing copper ores were discovered in the Congo at the Katanga Mine of the Union Minière. Extraction of cobalt began in 1924 and two years later Katanga became the leading producer, a position which she has easily maintained ever since.

With the opening up of the copper deposits in Northern

Rhodesia in 1930 cobalt was found at Nkana, Rhokana Corporation commencing production in 1933. Production in the United States dates from after World War Two, the Calera Mining Co., Idaho, commencing production in 1952 and the St. Louis Smelting & Refining division of the National Lead Company, Missouri, in 1955, both companies recovering cobalt by pressure leaching. The following table shows the output of the leading producing countries:

WORLD PRODUCTION OF COBALT (tons)

	1936	1940	1960
Congo	710	2,656	9,061
Canada	403	397	1,725
U.S.A.	—	64	785
Rhodesia	461	1,348	1,929
Germany	—	—	1,640
Morocco	371	364	1,475
Others	214	271	85
Total	2,159	5,100	16,700

PRODUCTION OF SMALT

Up to World War One, most cobalt was used by the glass and ceramic industries in the form of smalt for colouring and glazing glass, china, pottery and porcelain. The ores used for the production of smalt were smaltite, a cobalt arsenide ($CoAS_2$) containing when pure 28% cobalt, and cobaltite or cobalt glance (CoAsS) containing when pure 35·5% Co. Usually a portion of the cobalt is replaced by nickel and iron. Asbolite or wad is a cobaltiferous mixture of manganese and iron oxides, the cobalt content varying from 3 to 20%. It is the principal cobalt-bearing mineral in the New Caledonia deposits, containing 3 to 5% cobalt.

The ore was first roasted in reverberatory furnaces in order to convert the cobalt to the sesquioxide (Cr_2O_3). If much iron or nickel was present, the ore was not dead roasted, a certain amount of arsenide being deliberately left undecomposed in order to form a speiss (arsenide) during the subsequent smelting operation. On the other hand, if the roasting was terminated too soon cobalt remained combined with the arsenic and passed into the speiss and was therefore lost to the process.

The roasted ore was mixed with finely ground quartz (silica) and potash and placed in fireclay crucibles. Smelting was carried out in dome-shaped furnaces, the crucible being placed on tiles above the hearth. During the fusion, the sesquioxide, silica and potash combine to form a glass whilst the iron, nickel and any copper united with the arsenic to form speiss. While fusing the mass was stirred from time to time through the working doors located in the side of the furnace. On being allowed to stand, the molten mass separated into two layers, the speiss at the bottom, the cobalt glass at the top. The operation took from 10 to 15 hours. The glass in the crucible was then poured off, wet-ground in mills and sent through a series of washing vats in which the ground material was deposited according to size, the powder then being collected and dried. If left too long in contact with water the smalt lost its fine blue colour and became a grey-blue to dirty green. A typical smalt assayed 70% SiO_2, 6·5% CoO, 21·4% K_2O and 1·4% FeO.

PRODUCTION OF METAL

Production of cobalt is notable for the complexity and diversity of the operations employed. Pyrometallurgical, hydrometallurgical and electrometallurgical processes are all used to extract the metal.

The greater part of present-day cobalt production comes from complex-cobalt-copper and cobalt-nickel ores. The metal content of the ore varies within wide limits, but in general contains 0·1–0·5% cobalt. As a result, recovery operations are preceded by mineral dressing operations, generally flotation which concentrates the metal into a product containing up to 5% cobalt. The metallurgical process then adopted depends on the nature of the cobalt-bearing minerals (sulphides or oxides) and accompanying gangue material (basic or siliceous). Pyrometallurgical methods are used with both oxidized and sulphide minerals for (1) reduction to metal or cobalt-bearing alloy, (2) reduction to matte or speiss, (3) chloridizing or sulphatizing roasting. Cobalt is also isolated in the form of a high purity hydroxide or carbonate by hydrometallurgical methods, pyrometallurgical treatments subsequently being used for the production of cobalt oxide and reduction to metal.

Hydrometallurgical methods are also applied to both oxidized and sulphide minerals involving leaching, precipitation, separation and reduction to metal.

Electrometallurgical methods are used in the electro-recovery of cobalt metal and in the refining of metal produced by pyro-metallurgical and hydrometallurgical processes.

Arsenical Ores. One of the first smelting operations on a commercial scale was achieved in 1908 by the Deloro Smelting and Refining Company from ores derived from the Cobalt district in Ontario. The rise of the Canadian cobalt in this district has been closely related to the development of this silver producing area. The chief minerals are native silver associated with smaltite, niccolite (NiAs), cobaltite and a variety of other minerals containing arsenic, sulphur and bismuth.

The crushed ore was charged to a blast furnace with coke, limestone and iron scrap. Reduction at a temperature of 1100°C. gave rise to four products, crude silver bullion, speiss, matte and slag, which were tapped from the furnace and cast into pots. Gases, fume and dust were directed to a dust catcher where most of the heavy flue dust settled out, the fumes being blown by a fan into a baghouse where crude arsenic (As_2O_3) was recovered. On cooling, the contents of the pots settled out with the lighter matte on top, the speiss next and the heavy bullion at the bottom. When cold the layers were broken apart and separated by hand. The speiss (which contained most of the cobalt and assayed 20% Co, 12% Ni, 23% As, 18% Fe, 2% Cu and about 2% Ag) was roasted in Bruckner rotary furnaces to eliminate as much arsenic as possible. The calcine was then mixed with concentrated sulphuric acid to produce soluble cobaltous sulphate ($Co + H_2SO_4 = CoSO_4 + H_2$) the sul-phated speiss then being agitated with water, most of the copper, iron, cobalt, nickel and arsenic being dissolved. Silver sulphate, being insoluble in cold water, was filtered off and smelted to a crude silver bullion. The filtrate was treated with ground limestone to reduce the acidity, iron and arsenic being precipitated, and the solution then being freed of copper by addition of a small charge of iron powder. The filtered solution free from copper, containing 10–20 g/l of cobalt, was then treated with sodium hypochlorite which precipitated the cobalt as cobaltic hydroxide $(Co(OH)_3)$. This was then collected and heated to 720° C. converted to the black oxide (Co_3O_4) and reduced by heating at 1200° C. with charcoal in an electric furnace yielding a metal of approximately 97% purity. A similar procedure is followed in Morocco for recovery of cobalt from arsenical ores.

Union Minière du Haut-Katanga. Owing to the variation in cobalt content and the type of mineralization several distinct processes are operated by this concern. In general the following classes of ore are treated separately:

1. Copper oxide ore containing minor amounts of cobalt. The ore contains 5–6% copper, 0·2% cobalt and is bulk floated to 26% copper and 1% cobalt. The concentrate is then treated by leaching and selective electrolysis for recovery of copper and cobalt.

2. Cobalt copper oxide ore containing approximately equal quantities of the two metal constituents. A typical copper cobalt oxide feed contains 2·3% Cu and 2·0% Co, the flotation concentrate assaying 10% Cu and 8% Co. After sintering, the cobalt-copper oxide concentrate is treated by electric reduction smelting.

3. Mixed ores containing both cobalt and copper oxides and cobalt and copper sulphides. Selection flotation produces a copper concentrate assaying 46% Cu, 0·4–2·5% Co and is treated in fluo solid roasters before leaching. The oxide concentrate containing 20% Cu, 1·7% Co is either treated by hydrometallurgical processing or sintered and smelted in electric furnaces.

TREATMENT OF COBALTIFEROUS COPPER OXIDE CONCENTRATES

(1) At the Jadotville smelter the concentrates are treated with dilute sulphuric acid containing as reducing agent ferrous sulphate furnished by returned electrolyte from copper electrolysis. Both cobalt and copper go into solution as sulphides:

$$2FeSO_4 + Co_2O_3 + H_2SO_4 = 2CoSO_4 + Fe_2(SO_4)_3 + 3H_2O$$
$$Fe_2(SO_4)_3 + Cu = 2FeSO_4 + CuSO_4$$

After counter-current decantation to effect separation of the pregnant copper and cobalt solution, excess iron which is detrimental to current efficiency in electrolysis is precipitated by adjustment of pH to 2–2·5 by addition of copper hydrate. Electrolysis in two stages at high current densities lowers the copper content of the electrolyte from 20 g/l to about 1 g/l, the last traces of copper being removed by cementation by passage over metallic cobalt granules. The copper-free solution is treated with milk of lime which yields a precipitate containing cobalt. The filtered cobalt hydrate is repulped in water and forms the feed to the electrolytic tank house. Electric deposition of cobalt is only possible in a neutral solution achieved by electrolysis in a pulp containing excess of cobalt hydrate in suspension (up to 70 g/l) the hydrate acting as a neutralizing agent for acid generated at the lead anodes. Power consumption is of the order of 6·5 kwH per kilogram of cobalt at a current density of 50 amps/sq. ft. The cathodic deposit (92–94% Co) is melted in an electric furnace, the chief impurity zinc (1–2·5%) being volatilized, the molten cobalt then being granulated by pouring into water. The usual composition is Co 99·3%, Ni 0·4%, Fe 0·2%, Zn 0·01% and Cu 0·01%.

(2) Sulphide copper cobalt concentrates. The sulphide concentrate containing 46% Cu, 0·4–2·5% Co, 12% S and 1–2% Fe is treated in a fluosolid roaster (page 62) at a temperature of 670° C., whereby both the copper and cobalt are converted to sulphates, the greater part of the iron being left as insoluble oxide. The sulphate calcine is leached with sulphuric acid and joins the main production stream in (1) following the same steps as for the recovery from oxidized ores.

(3) High grade cobalt concentrates. Oxidized copper-cobalt concentrate containing 10% Cu and 8% Co are mixed with finely powdered coal and sintered, the sinter being incorporated with coal and lime and charged to electric furnaces operating at a voltage of 120 and a current of 5,000 amps.

By melting the charge in a reducing atmosphere, cobalt, copper and iron are obtained in the metallic state. Two alloys are produced which on tapping separate in the ladle by difference in density. The heavy 'red alloy' is rich in copper and poor in cobalt (89% Cu and 4·5% Co) and is refined to copper at the copper smelter. The lighter 'white alloy' containing approximately 42% Co, 15% Cu and 34% Fe is cast into ingots and sent to Belgium for recovery of Co and Cu by dissolution in dilute sulphuric acid, cobalt only going into solution. It is precipitated as carbonate and reduced to metal.

The percentage contribution of Katanga to cobalt production has varied from 50–60% making this area by far the most important producer. Capacity at about 10,000 tons per year can be extended to 13,000 tons, reserves of cobalt ore being very high.

Rhokana Corporation Ltd., N. Rhodesia. Cobalt occurs as a minor constituent (0·10%) of the Rhokana ores which are mined essentially for the copper content. In the flotation circuit the bulk of copper is removed after which a cobalt concentrate is recovered assaying 2·75% Co and 31% Cu. Prior to the last war this was smelted in a reverberatory furnace,* the cobalt entering the matte along with the copper. During the subsequent conversion the cobalt was slagged off with the iron, mixed with coke and smelted in an electric furnace when the cobalt was recovered in the form of an alloy assaying 37% Co, 12% Cu and 50% Fe. The alloy was granulated and shipped overseas for recovery of cobalt by wet processing. The overall recovery of cobalt from ore to alloy was extremely low being about 14%. After the war the cobalt grade declined, an increasing proportion of chalcopyrite and pyrite in the ore considerably raised the iron to cobalt ratio, and as a consequence cobalt recovery steadily decreased until it became increasingly difficult to produce alloy of a grade acceptable to the refiners. As a further increase in this ratio was to be expected from the ores to be treated, investigations were put in hand to devise an alternative process. Roasting and leaching of the cobalt concentrate was finally selected, followed by electrolysis, output of electrolytic cobalt commencing in 1953.

The cobalt concentrate containing 25% Cu, 17% Fe and 3–5% Co is given a sulphatizing roast with the object of converting the cobalt oxide to sulphate leaving most of the copper and iron as insoluble oxides:

$$CuSO_4 + CoO = CoSO_4 + CuO$$

$$CoO + SO_3 = CoSO_4$$

* H. L. Talbot and H. N. Hepker. *Production of Electrolytic Cobalt.* Bulletin of the Inst. M.M., September 1949.

The roasted concentrate is pulped with water, agitated at 80° C. and filtered, the residue containing most of the copper passing to the smelter for incorporation in the reverberatory furnace charge. The filtrate containing 10 g/l Co, 8 g/l Cu, and 0·2 g/l Fe is treated with milk of lime under vigorous air agitation at a pH of 6·5 to precipitate most of the copper and iron, the last traces of copper being removed from the cobalt solution by passage over granulated metallic cobalt. Addition of milk of lime to the clear solution precipitates cobalt as the hydroxide which is filtered off, the cake being then dissolved in spent (acid) electrolyte to give a pH of 6, the solution for electrolysis containing 20–30 g/l cobalt.

Electrolysis is carried out in a similar manner to that practised at Katanga with lead anodes and mild steel cathodes at a current density of 17 amp/sq. ft. A five-day cathode cycle results in the deposition of about 50 pounds of metal per cathode. The excess of acid generated during electrolysis is neutralized by addition of cobalt hydroxide. Cathode deposits are melted in an electric furnace and granulated, the purity of the metal being about 99·8%.

Pressure Leach Treatment. The technique of leaching under pressure has in recent years been applied to recovery of cobalt. Sherritt Gordon Mines Ltd., in Canada, Calera Mining Co., and the National Lead Co., in the United States, all treat cobalt-bearing ores by pressure leaching and hydrogen reduction processes (page 212). At the Garfield plant of the Calera Mining Co., a cobalt concentrate assaying 17% Co, 20% Fe, 24% As and 30% S, is fed to an autoclave operating at 200° C. and 600 psi. An exothermic oxidation reaction converts the metals to sulphates, the iron and arsenic combining to form insoluble iron arsenates. The slurry from the autoclave is filtered, the resulting cobalt solution being purified by lime additions, pH adjustment and cementation to remove iron, arsenic and copper. In the hydrogen reduction operation, the solution is made alkaline with ammonia and passed to autoclaves into which hydrogen gas is injected to give a pressure of 800 psi to yield cobalt in the form of a fine powder.

This brief review of the principles applied in the recovery and refining of cobalt give a rough idea of the complexity of the methods and the equipment used. The plants are designed and built in most cases with a view to producing one principal metal, such as copper or nickel, cobalt being a by-product. However, that part of the installation which is required for the separation and refining of cobalt is large and costly; for example, in the joint copper and cobalt recovery plants, the investments pertaining exclusively to cobalt may reach 35% of the total, whereas cobalt product represents only about 3 or 4% tonnage-wise of the marketable metals.

In the evolution of methods of production pyrometallurgical methods are being increasingly replaced by hydrometallurgical and electrolytic processes. Among such changes the reduction to metallic form by electrolysis was applied for the first time on an industrial scale in Katanga in 1945. Since that time many other producers have applied this process in one form or another. Elsewhere reduction by hydrogen under pressure has resulted in a metallurgical process entirely free from pyrometallurgical operation. The advantages of these hydrometallurgical processes is that they lend themselves to continuous operation and to automation and as a consequence recovery yields are increased, reduction in man power is effected and the grade of the product is improved.

Uses. The following table concerns America only, but can be regarded as a fair indication of world trends:

COBALT CONSUMPTION (IN THE UNITED STATES)

Uses	1961 %
High-temperature alloys	24·6
Cutting and wear-resisting alloys	2·7
Magnet materials	25·0
Hardfacing rods	5·8
Cemented carbides	2·8
High-speed steels	2·1
Other tool steels	0·4
Other alloy steels	5·4
Non-ferrous alloys	1·5
Other metallic uses	7·3
Total Metallic	77·6

COBALT CONSUMPTION (IN THE UNITED STATES)—*contd.*

Uses	1961 %
Salts and driers	13·0
Ground-coat frit	5·5
Pigments	1·7
Other non-metallic uses	2·2
Total Non-Metallic	22·4

Although the use of cobalt as a colourant in the manufacture of glass and ceramics dates from ancient times, virtually no metal was produced until about 40 years ago. In 1875 E. Haynes in the United States began an investigation into corrosion-resistant alloys, and by 1899 had developed an alloy containing 75% cobalt and 25% chromium usable as a metal-working tool material possessing unusual corrosion resistance, toughness and retaining its hardness at elevated temperatures. He called these alloys stellite from the Latin word *stella*, meaning star. Further patents were granted in 1913 to cobalt-chromium-tungsten alloys used for high-speed lathe cutting and these became important in machining munitions produced in World War One.

The metal cobalt is one of three elements that are ferromagnetic at room temperature. The first cobalt steel for permanent magnets was developed in 1916 by K. Honda, a Japanese scientist. It contained 5% tungsten, 6% chromium, 35% cobalt, remainder iron, and was many times more effective than the chromium or tungsten magnetic material previously used. In 1930 an improved magnetic material was discovered by another Japanese scientist, Professor T. Mishima. This alloy at first contained 12% aluminium and 25% nickel, balance iron, but following this discovery it was found that cobalt additions improved considerably the magnetic properties of the alloy. This new series of permanent magnet materials became known as the Alnico's being based on the composition 5–11% Al, 12–25% Ni, 5–25% Co, balance iron. These alloys cannot be forged or machined and are now usually worked by powder metallurgical techniques.

They are mainly used in the manufacture of small motors,

generators, magnetos, loud speakers, television tubes, radar equipment, etc.

One of the largest uses of cobalt today is in alloys designed for high temperatures, their use having increased rapidly as a result of the development of the gas turbine and jet engine. In 1909 Tammann suggested a cobalt-chromium alloy containing iron or nickel for steam and gas turbines. The cobalt chromium base alloy known as Vitallium developed in 1936 for the precision casting of dentures was found to possess unusually high temperature strength at temperatures of the order of 815° C. and was used during World War Two for the super-chargers of aircraft engines. As blade material, however, brittle failures in the leading edges of the blades were encountered. With the addition of small amounts of nickel to improve resistance against oxidation and brittleness, the material was able to meet specifications.

In Great Britain, the development and production during and after the last war of a series of alloys based on the nickel chromium system has contributed significantly to the progress made with such engines. The substitution of 20% cobalt for nickel with small additions of titanium and aluminium was subsequently shown to improve considerably the heat-resisting properties.

BIBLIOGRAPHY

1. *Cobalt*. Monograph edited by Centre D'Information du Cobalt, 1960. Brussels, Belgium.
2. *Cobalt, Its Chemistry, Metallurgy and Uses*. Edited by R. S. Young. Reinhold Publishing Corp., New York.
3. WILKINSON, W. D. and MURPHY, W. F., *Nuclear Reactor Metallurgy*, 1958. D. Van Nostrand Co. Inc., New York.
4. IRANI, M. C., *Processes for Extraction of Uranium from Ores*. Deco Trefoil, January–February 1956.
5. ROBINSON, R. E., *Ion Exchange in Uranium Production*. Optima, December 1956. Anglo American Corp. of S.A.
6. EVERHART, J. L., *Titanium and Titanium Alloys*, 1954. Reinhold Publishing Co., New York.
7. Production of New Metals, *The Engineer*, December 1959.
8. PFANN, W. G., *Zone Melting*, 1958. J. Wiley & Sons Inc., New York.

CHAPTER SEVEN

THE PRECIOUS METALS

UNDER this title are included the high-price metals: gold, silver and platinum; palladium, iridium, rhodium, ruthenium and osmium; the last six being referred to collectively as the platinum metals, for they are usually found in nature associated with platinum.

The untarnishable nature of gold and to a smaller extent silver, led to their being amongst the first metals to be discovered and recognized, gold being well known to the Egyptians 8,000 years ago. Of this metal Dr. Charles Seltman says 'Found in the earth almost unalloyed, untarnishable, its purity capable of test by fire whence it emerged unaltered, everlasting, immutable, it became a kind of symbol of immortality for which mankind forever hankered.'

Its desirability for decorative purposes and as a medium of exchange has been long appreciated, especially in the near Eastern lands whence our civilization began. Since earliest times, desire for its acquisition has been a major cause of wars, and in more recent times the search for the metal has exerted a powerful influence on expanding the frontiers of nations. The opening of the West in the United States in 1849 following the discovery of gold in California, the spectacular gold discoveries in Victoria, Australia, in 1851, and the discovery of the Witwatersrand goldfields in 1886 are all examples of communities gold has helped to create.

Gold commonly occurs in veins or lodes associated with quartz which usually forms the gangue material.

Disintegration of the auriferous veins by natural agencies such as water, leads to the formation of sands and gravels known as placers. Gold also occurs in sea water in amounts from $0 \cdot 03$–1 gm per ton, but all attempts to extract it at a profit have so far failed.

At the present time gold comes mainly from lode deposits,

this being the form of the deposit in South Africa, the chief gold-producing country. Placer gold, however, before the discovery of the Witwatersrand in 1886, was responsible for much of the world's supply. During this period the gold diggings of California, the Klondike and Alaska reached their zenith.

Placer Gold.   Treatment was usually by gravity methods in more or less simple appliances, such as the pan, cradle, sluices, and by dredging. The gold pan, which is more a testing apparatus than a recovery tool, is not unlike a frying pan with its flat base, 12–24 in. in diameter, and sloping sides about 4–5 in. deep. In use the gravel or sand is placed in the pan, immersed in water and given a semicircular motion keeping the pan horizontal, the effect of which is to bring the lighter material to the surface, the heavier grains sinking to the bottom. The pan is then tilted and gently rocked from side to side in order to wash away the lighter material. Continual rocking and tilting result in the greater bulk of the ore particles being washed over the edge of the pan leaving the gold at the bottom of the pan. The oldest and most primitive forms of recovery equipment depend for their action on the fact that in a flowing stream of water the heavier particles carried by water tend to settle and collect against any obstacle in the bed of the stream. This is the principle of the rocker, Long Tom and other forms of riffled sluices in which the material is washed down an inclined trough or launder, subjected to a rocking motion, the coarse particles of gold being caught against strips of wood (riffles) nailed at intervals across the bed of the launder.

Larger scale operations are by means of a rocker, which has a greater capacity than a pan. This appliance, which was widely employed by the forty-niners, resembles a child's cradle in appearance, and consists of a wooden trough 4–5 ft. long, sloped at about 2 inches per foot and mounted on two rockers. Material is shovelled on a perforated screen at the head of the rocker and washed through by water, any rock or boulder being thrown out. The undersize material is caused to progress along the sloping floor by rocking the cradle any gold being caught by transverse riffles, the lighter gangue material being washed out of the tail end of the machine. Coarse gold is readily caught, but such an appliance loses much fine gold, which can to some extent be remedied by placing mercury in the riffles which aids in catching the gold.

Two men working the rocker, one feeding, the other rocking, could account for 4–5 cu. yds. of gravel per 12-hour day. Many thousands of rockers together with a similar appliance known as a Long Tom were used by small workers in the 1849 Californian gold rush, and later in Alaska and the Klondike.

In the period 1900–15 dredging became the most important branch of placer working. Originating in New Zealand in the 1860s, the dredge consists of a large boat or raft 150–200 ft. long, on which are mounted at the bows an endless chain of buckets which dig up and elevate the gravel delivering it to the recovery plant, consisting of jigs and tables. The dredge floats in natural water or in self-excavated ponds which it enlarges ahead as it works, tailings being deposited astern. Digging depths average 40–50 ft. below water-level.

Amalgamation. The process of amalgamation depends on the fact that when metallic gold (and silver) come into contact with mercury they combine to form an alloy known as amalgam.* The amalgam is generally whitish in colour and is solid, pasty or liquid according to the proportion of mercury it contains.

Amalgamation is carried out in one of the two following ways:

1. In plate amalgamation the crushed ore, together with water, is brought into contact with surfaces coated with a thin layer of mercury.

2. In pan amalgamation the ore and mercury are ground together in an iron pan with sufficient water to form a paste, chemicals being added as necessary to assist the reaction.

During the days of the Californian gold rush and until the introduction of the cyanide process at the turn of the last century, plate amalgamation was the conventional method, pan amalgamation being used principally when the material treated was essentially ores of silver. Plate amalgamation as carried out by the forty-niners consisted in crushing the ore in presence of water in stamp batteries until the ore particles were fine enough to pass through the battery screens. Mercury together with

* Gold and silver are not the only metals which combine with mercury; zinc, lead, bismuth and sodium also amalgamate.

sheets of copper coated with mercury were placed inside the mortar box (known as inside amalgamation), amalgamation plates being also placed on sloping tables outside and in front of the screens, the pulp flowing over them as it left the batteries, the gold particles having a high specific gravity settling through the flowing stream to be caught by the mercury on the plates.

The plates used for amalgamation consist of pure copper $\frac{1}{16}-\frac{3}{8}$ in. thickness, width of plates being 48, 52 and 54 in. with standard lengths of 4–8 ft.; 2–4 sq. ft. of surface plate per ton of ore were usually necessary. The plates are amalgamated by first thoroughly cleaning with emery paper or fine sand until quite bright, mercury then being rubbed in with the aid of salammoniac. The total amount of mercury used varies with the richness of the ore treated, 1–2 oz. per oz. of gold being perhaps an average figure. The gold amalgam which accumulates on the table is periodically removed by means of a rubber scraper, and placed in a canvas or chamois leather bag; the excess of mercury is squeezed out, the solid residue remaining in the bag being heated in a retort to distill off the remaining mercury, leaving behind the gold. The disadvantages associated with the use of amalgamated plates, namely the 'lock-up' of gold, the possibility of loss by theft and of mercurial poisoning among the workmen, led in many of the gold-producing plants to substitution of corduroy cloth for plate amalgamation. Corduroy is fixed to a table with its ribbing at right angles to the flow of pulp, the gold particles lodging in the depression, the lighter gangue material flowing away with the water.

The legend of the Golden Fleece which was found at Colchis, the port for the rich Caucasian mines, may well reflect the prosaic use of fleece to catch the particles of gold, very much as corduroy is still used in gold plants today.

Amalgamation in cast iron pans was operated to a considerable extent for the treatment of ores which did not yield their gold to amalgamated plates or which contained much silver. The method of treatment was similar to that employed for silver ores (page 285). The method is now obsolete.

Cyanide Process. That gold was dissolved by a solution of a potassium cyanide was known in the early part of the 19th century. A. Parks applied for a patent in 1840 in which separation of gold from its ore was to be effected by digestion with a 3–6% solution of cyanide of potassium, for 3 days at 150–180° F. It was then proposed to recover the gold by fusing the residue obtained by evaporation of the solution. The process was obviously uneconomic involving as it did large losses of cyanide.

L. Elsner, a German chemist, published in 1846 a paper incorporating the basic idea of cyaniding, but apparently failed to recognize its significance and no practical use was made of his findings.

A patent which had much more bearing on the subject was that issued to an American, J. W. Simpson, in 1884. In this patent the crushed ore was agitated with a 2½% solution of cyanide of potassium, The solid material was then allowed to settle, and a plate of zinc suspended in the clear liquid, which causes the gold dissolved in the solution to be precipitated thereon, from which it could be removed by scraping or by dissolving the zinc in sulphuric or hydrochloric acid.

It would seem in the light of modern knowledge that although the cyanide strength was excessive and that the use of zinc in the form of plate would result in insufficient precipitation; nevertheless, if the process had been introduced on a large scale it could have been so modified as to achieve success. As it was it was left to the enterprise and skill of J. S. MacArthur, a metallurgical chemist, and R. W. and W. Forrest, Doctors of Medicine of Glasgow, to make a practical success of the cyanide process for the extraction of gold. An interesting account of the research leading up to the development of the process was given by J. S. MacArthur in a paper read before the Scottish Section of the Society of Chemical Industry in March 1905.

The process first under investigation depended on the solvent action of chlorine generated electrolytically in an alkaline solution. It was found, however, that such solvents attacked the base metals in the ore in preference to the gold. The efforts of the researchers were therefore directed to finding a gold solvent that would be inert to base metals. A variety of solvents were tried—ferric chloride, ferric bromide, etc., and in November, the effect of sodium cyanide on the tailing of an Indian gold mine was tested and as usual the solution was treated with H_2S to detect the presence of gold. Obtaining no reaction, the next solvent on the list was experimented with. Eleven months later, having occasion to test again a cyanide solution which was known to contain gold with H_2S, and obtaining a negative response, they were appraised of the fact that H_2S did not invariably precipitate gold from cyanide solution. It was then realized that the experiment carried out nearly a year before might have been successful without success being recognized. A sample of concentrate from a Californian mine was treated, the residue being examined and not the solution, a high percentage of extraction being found. The original residues from the Indian gold ores were then re-examined, and it was found in that case too the gold had been transferred to the cyanide solution. There was then no doubt of the discovery; a provisional specification was lodged in 1887 and a complete one taken out in 1888 entitled 'Improvements in obtaining gold and silver from ores and other compounds'.

The following extracts from the specification will indicate its essential points:

1. Nature of the Solvent. The ore in a powdered condition is treated with a solution containing cyanogen or cyanide (such as cyanide of potassium, sodium or of calcium) or other substance or compound containing or yielding cyanogen.

2. Use of dilute solution. In practice the best results are obtained with a very dilute solution, such solution having a selective action in dissolving the gold or silver in preference to the baser metals.

Apart from a statement that 'the solution drawn off from the ore may be treated in any suitable way as, for example, with zinc for recovering gold and silver' no specific means of precipitation is described, but in the 1888 patent a method of precipitation is described in which the separated gold solution is made to pass a mass of zinc in a state of fine division such as shavings, thin strips or grains. A. P. Price, in 1883, and J. W. Simpson, in 1884, had both previously used sheet zinc as a precipitating agent, but MacArthur and Forrest evidently found that sheet zinc is not so efficient as zinc in filament form.

These two discoveries—namely that the solution of gold and silver can be effected by a dilute solvent of cyanide and that the gold can be recovered by precipitation on finely divided zinc—revolutionized the metallurgy of gold. Between the date of the discovery in 1886 and 1896, world production of gold was doubled, and again doubled in the following decade. In 1889 the world consumption of cyanide did not exceed 50 tons per annum, whereas in 1905 the consumption was nearly 10,000 tons per annum, of which the South African goldfields alone took a third.

It would seem that the first works operating on a commercial scale was at the Crown Mine, New Zealand, in 1889. In 1890, the process was started on a commercial basis by the Gold Recovery Syndicate representing the Cassel Company (the owners of the MacArthur–Forrest patents) in South Africa. A contract was entered into to treat 10,000 tons of tailings from the Robinson Gold Mines, Johannesburg. It is interesting to note that (in these days of 4 dwt. gold values) any tailing of the gold value of 8 dwt. or under was not to be paid for. Throughout the contract a recovery of 75% was obtained, and from that time onwards the cyanide process was established and quickly adopted by gold mines in all parts of the world. The first cyanide plants in the U.S.A. were erected in 1891, one at the Consolidated Mercury in Utah and the other in Calumet, California. The method initially used by the Gold Recovery Syndicate consisted in agitating the tailings in a tank in contact with cyanide solution and subsequent filtration on a suction filter, the gold in the filtrate being recovered by precipitation with zinc shavings. It was considered that the intimate contact secured by this method ensured a more rapid and a superior dissolution of gold. Largely, however, because of the higher power consumption and because of the tendency of the coarser particles to settle out and escape treatment, the method was abandoned in favour of the more simple method of leaching by percolation. As will be shown later the agitation method has now come back to favour and is the standard process.

The term percolation signifies the passage of a liquid through a mass of stationary material contained in a tank. In the early days of the process square wooden tanks were used, but as these caused heavy losses by leakage, circulatory tanks of steel or of wood tightened by hoops were introduced. The bottom of the tank enclosed a filter frame composed of a lattice work of crossed wooden slats on which rested a filtering medium composed of hessian or coconut matting. The ordinary sized 25 ft. diameter × 5 ft. high tank holds 100 tons of tailings or ground ore, and a tank 50 ft. in diameter by 8 ft. high, 600 tons.

The leaching tanks were filled with tailings or crushed ore from side-tipping cars running on an overhead railway. When full, cyanide solution of 0·10% strength was run on top of the charge until completely covered and left standing for 24 hours or so, then drained off to be followed by a weak solution to displace any of the strong solution and to dissolve any gold not extracted by the strong solution. The liquor used for this purpose was generally that which passed out of the precipitation unit. Finally, the charge was treated with a water wash. The spent residue was then discharged through openings in the bottom of the tank into trucks below, or where facilities existed for the purpose the residue was discharged by sluicing with water.

The total amount of solution used would be about twice the weight of material, the total time of treatment being about four days.

The zinc boxes in which precipitation of gold from solution was effected consisted of a long narrow steel trough 12–20 ft. long, divided by transverse partitions into a number of compartments, the bottoms being composed of wire screening on which rests the zinc. The partitions were so arranged that the solution rose up through the column of zinc and flowed down through the intervening compartment. Recovery of the precipitated gold took place once or twice a month when the flow of solution was shut off and, commencing with the head compartment, a jet of water was directed on the zinc for the purpose of loosening the gold deposit adhering to the zinc, and washing through the wire screening, a plug being then opened at the bottom of the compartment and the sludge run out into a clean-up tank. The zinc in the compartment was then removed, washed free of adhering gold and replaced.

The remaining compartments which contained sufficient gold were treated in the same manner, the zinc being replaced in the compartments, any shortage being made good by addition of fresh zinc and the box put back into operation.

The precipitate in the clean-up tank was allowed to settle, the clear solution siphoned off and the precipitate treated with 10% sulphuric acid to dissolve the zinc. When all the zinc was dissolved, the tank was filled with water, agitated, allowed to settle and liquid siphoned off. The gold slime was then removed, dried, mixed with borax, sand and soda ash, and smelted in a graphite crucible, the resulting bullion being run into a mould.

Improvements in Sand Leaching. In its essentials the above was the method practised in the early days on the Rand goldfields and elsewhere, and because of its simplicity is still occasionally carried on by small operators in remote workings. The method is, however, inadequate for the treatment of large tonnages and for sands containing much slime, and in succeeding years more complicated arrangements both in plant and practice were introduced.

At an early stage it was found that the treatment of the crushed material was often unsatisfactory in that instead of percolating evenly throughout the whole charge channels were formed, i.e. in the coarse sandy areas, wherever there was least resistance. On the other hand, the fine slimy material formed areas practically impervious to the solution. Hence many areas of the charge did not come into adequate contact with the cyanide solution and the extraction of gold suffered accordingly.

A rough separation of the percolable material was originally made by running the crushed ore into pits where the sandy portion was deposited, the slime overflowing to waste. Separation of slime was also effected in the Spitzkasten (introduced by Rittinger about the middle of the 19th century) consisting simply of a pointed pyramidal box, and its modification Spitzlutten in which an upward current of water was introduced from the bottom or apex of the box. The pulp fed in at the top passes downwards and, on meeting the ascending stream of water, the lighter particles are carried up overflowing at the top of the box, the coarser and heavier particles falling through the outlet at the bottom. Crude classifiers in the form of a hollow inverted cone made of sheet iron were also in use for separating sand from slime. A typical cone measuring 8 ft. diameter by 8 ft. in length was fitted with a central down pipe conveying water which impinged on a circular disc situated several inches above the apex of the cone, the effect being to create a circular sheet of water through which the sands had to pass before escaping. By using a number of these cones in series a clean sand product was obtained.

For many years there was no known method for the economic treatment of the slime which amounted to roughly 30% by weight of the charge, and the operators avoided as much as possible the formation of slime in milling. In time, however, methods were developed, and it is in the treatment of slimes that the chief advances in the cyanide process were made in succeeding years. One of the first methods for handling gold slime, originated on the Rand in 1894 by J. Williams, was known as the decantation process. The method consists of successive agitations with cyanide solution, settlement and decantation, this cycle of operation being repeated two or three times. Lime was added to assist settlement. The clear supernatant solution which gathered above the settled slime was removed by a decanter consisting of a hinged pipe that passed through the side of the tank, with its inlet kept just below the surface of the clear solution by means of a hinge and a float. The decantation process for slime treatment suffered, however, from many defects. High extraction could be obtained only by a large number of successive decantations involving large quantities of solution which rendered the process prohibitive where water was scarce. Further, it was a batch-operated method, and like all such methods was time and labour consuming, especially when slime with indifferent settling characteristics was encountered.

These objections were to some extent met in Australia and elsewhere by the employment of the plate and frame type filter press, the pulp being forced through the press by pumping, a pressure of up to 80 lb. per sq. in. being required, the gold solution passing through the cloths and out through the taps, the slime forming a solid layer on the surface of each filter cloth. When filled, compressed air is admitted which displaces solution from the slime cake. Wash solution is then applied and finally air is again admitted to remove any excess moisture, the press opened and the cake discharged. The advantages of this system are that the greater part of the dissolved gold values may be extracted in one operation in a short time and with a much smaller volume of solution than is possible by decantation. As the filter leaves have to be taken apart for the removal of the slime cake the labour costs are high. To obviate this expense suction leaf filters employing a vacuum were developed, the Moore and the Butters in 1903, later in 1907 the rotary drum filter, the invention of E. L. Oliver, and then the American disc filter. In all these types a vacuum is applied to form a cake on a filter cloth, the cake then being dislodged after washing by cutting off the vacuum and applying compressed air. The rotary type of filter superseded the Moore and Butters leaf filter and is now used in all parts of the world.

Agitation of the slime pulp was effected by mechanical means aided by injection of compressed air for the purpose of introducing oxygen. Large shallow tanks were at first used, but it was found that by increasing the depth of the tank and giving it a coned bottom, mechanical agitation could be dispensed with, being displaced by air. This led to the development of the Brown agitator by F. C. Brown at the Komata Reef Gold Mine in New Zealand in 1902. The tank consisted of a tall vertical steel cylinder 30–50 ft. high and 10 to 15 ft. diameter with a conical bottom of 60 in. slope. Air was delivered through the apex of the cone to a central open air lift pipe at a pressure just sufficient to overcome the weight of the pulp column at the point of introduction, thus establishing a circulation of the slime, the air bubbles carrying the pulp with them up the pipe rising to the top and overflowing to the main body of the pulp in the tank. The Brown vat was extensively used in Mexico, especially in the silver district of Pachuca, and hence was known in the Americas as the Pachuca tank.

Finally, an agitator which combined both air and mechanical agitation was invented by J. V. N. Dorr at South Dakota in 1907.

From World War One the process of cyanidation had become well established and the forerunner of the present equipment and mode of operations had appeared. The introduction by Dorr in the period 1900–10 of the mechanical classifier, thickener and agitator and the appearance of more efficient means of treating and filtering the slime portion of the pulp tended to a situation in which all the ore was ground to slime, no separate treatment of sand and slime being provided for; and it was evident from an early date that the retention of sand leaching as a means of gold recovery was on the wane.

The all-sliming process of gold extraction is now practically universal in all the major gold-producing countries where free milling ores are concerned.

Multiple stage crushing became standard practice, using jaw or gyratory crushers for primary crushing with cone crushers as secondaries, the ore being reduced to — $\frac{1}{2}$ in. Stamp milling which was so long a feature of gold milling for the finer crushing of ore has now virtually ceased to exist (no stamp batteries have been erected on the Rand since 1918), having been superseded by the cone crusher. For grinding to the requisite degree of fineness, the ball mill or tube, or both, are employed in closed circuit with mechanical classifiers. Grinding in cyanide solution was first developed at the Crown Mine in New Zealand in 1897, but later abandoned; the losses via launders and the dissolved values in slimes outweighed the advantages gained by early contact of the cyanide solution with ore.

With the advantage of better equipment for treatment of slime, grinding in cyanide reappeared and is now standard practice. Use of zinc shavings has practically disappeared, having been replaced by the Merrill–Crowe zinc dust precipitation process, advantages being that with zinc dust a complete clean-up of gold takes place at the end of each month, whereas with zinc shavings an amount of gold approximating to one month's output is carried forward in the boxes. There is also greater security against theft and a saving in zinc and smelting costs.

In brief (Fig. 20), the ground pulp overflowing from the classifiers at about 20% solids is thickened (de-watered) to 50–60% solids in continuous rake-type thickeners and pumped to Pachuca tanks where it is agitated for 20–30 hours for solution of gold. Filtration is done on Oliver filters, the filter cake being repulped, refiltered and pumped to the slimes dam. Filtrate is clarified and then goes to deoxidation (vacuum) towers, addition of zinc dust precipitating gold which is collected in small filter presses and smelted.

Data relevant to gold extraction are as follows:

Head Values dwt. per ton	5–6
Mill feed, screen size (in.)	$\frac{1}{2}$
Primary classifier overflow—200 M %	80–90
Solids %	20–25
Thickener—Solids %	50–60
Cyanide Solution Strength % K.C.N.	0·01–0·05
Cyanide Consumption lb. per ton of ore	0·1–0·4
Lime solution strength % CaO	0·02–0·04
Lime consumption lb. per ton	1·5–4·0
Barren solution dwts. per ton	0·01
Zinc dust lb. per ton of solution	0·05

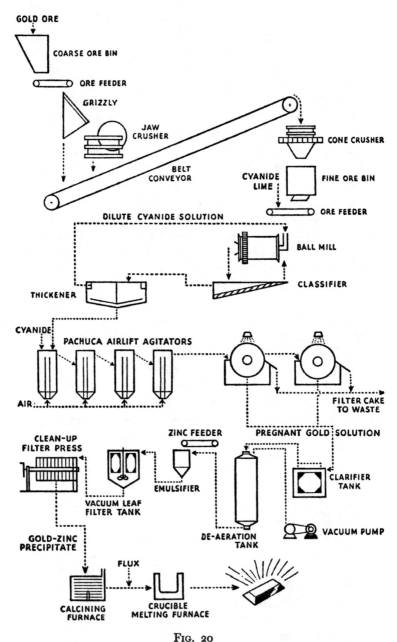

FIG. 20

GOLD RECOVERY BY 'ALL-SLIMING' CYANIDE PROCESS

Residue values dwts. per ton 　　　　　　0·3–0·5
Total Gold Recovery %　　　　　　　　　　90–95
Power consumption kW-hr. per ton
　　milled　　　　　　　　　　　　　　　20–30

REFINING

Gold bullion obtained from cyaniding and other processes always contains silver and frequently zinc, lead, copper and iron, the amount varying according to the nature of the impurities present in the ore and to pick-up by the cyanide. Thus a refractory gold ore may result in a bullion assaying only 500 fine whilst gold from amalgamation can assay as high as 980. In order to fit the metal for coinage, jewellery and industrial use, the metal must be refined, the fineness being raised to at least 995. It is true that the great bulk of the world's present output of gold is turned into bullion bars and stored indefinitely in vaults; but gold for this purpose must still be marketable as fine gold.

The separation of gold from silver to obtain each metal in a commercially pure state is technically known as 'parting'. The earliest methods of parting depended for their success on the use of acids which reacted selectively upon one or other of the metals.

The earliest separation of gold from silver and base metals was effected by the use of nitric acid. The method depends on the fact that silver is soluble and gold insoluble in nitric acid:

$$6Ag + 8 \, H \, NO_3 = 4H_2O + 6Ag \, NO_3 + 2NO$$

Nitric acid, however, cannot attack any alloy with a gold content in excess of 30%, silver having to be added to the alloy in order to allow it to part. Silver is recoverable from the solution by addition of salt, the insoluble chloride being precipitated

$$Ag \, NO_3 + NaCl = NaNO_3 + AgCl,$$

the silver chloride being then reduced to metal by means of scrap zinc or iron.

$$Fe + 2Ag \, Cl = FeCl_2 + 2Ag$$

Nitric acid for large-scale refining was abandoned in favour of sulphuric acid introduced about 1840 at the Royal Mint and elsewhere, the reasons being as follows: (1) high acid cost; (2) objectionable nitrous fumes; (3) tin, antimony and arsenic form compounds insoluble in nitric acid and hence contaminate the gold.

Silver dissolves in sulphuric acid according to the following equation:

$$2Ag + 2H_2SO_4 = Ag_2SO_4 + 2H_2O + SO_2$$

gold not being attacked. Rich bullion as before must be alloyed first with silver to reduce it to parting quality. Bullion in granulated form is added to the acid contained in cast-iron pots and heated to facilitate solution. When solution is complete, the clear solution is siphoned off into lead-lined tanks. On dilution first with cold acid and then cold water, silver sulphate is precipitated which is separated, washed and reduced with scrap iron:

$$Ag_2SO_4 + Fe = FeSO_4 + 2Ag$$

The precipitate is collected, dried and melted, giving a product of 996 fineness. The gold residue in the pot is treated with fresh acid to remove any remaining silver and base metals, washed with water, dried and fluxed with nitre and bone ash and cast into bars having a fineness of 990–995.

Because of the cheapness of the process it persisted for many years, but has now been abandoned by the large refiners in favour of chlorine and electrolytic methods.

Chlorine Gas. This method, which was first applied at the Sydney Mint in Australia in 1867 by Dr. F. B. Miller is now used for refining the whole output (70,000 oz. per day) of the South African goldmines and hence the bulk of the world's output. The process is based on the fact that chlorine readily combines with silver and any base metals present forming chlorides, gold remaining unaffected.

The process is carried out in clay pots which are generally placed in the furnace in 'guard' crucibles of plumbago. The pots have a capacity of 1,000 oz. of gold and are provided with a lid perforated with a hole through which a clay pipe for introducing the chlorine is inserted.

When the bullion is molten the chlorine is turned on, the

flow being adjusted so that no bubbling takes place, the chlorine being completely absorbed by the base metals, the majority forming volatile chlorides which are drawn off by fans into settling chambers and electrostatic precipitators.

The end point of chlorination is denoted by the appearance of dark brown fumes of gold chloride which give a characteristic stain on a piece of pipe stem. The clay pot is then removed from the furnace, the molten chlorides skimmed off, and the gold poured into moulds the bars assaying 996–997 fine. The time required for chlorination depends on the amount of impurities in the bullion and varies from 2–3 hours.

The chloride scum removed from the refined gold contains from 5–10% of gold and is separated by melting in plumbago crucibles with soda ash and carbon. Silver chloride is reduced, and the metallic silver in settling carries down the gold, the bullion of the reduced metal containing 40–50% of gold. The rest of the silver may be recovered by agitating the chlorides with brine, the base metal chlorides going into solution leaving silver chloride insoluble. This is reduced with iron in a dilute sulphuric acid solution.

Electrolytic Parting. Electrolytic methods are essentially wet methods and are based on acid processes. The advantages over acid parting are lower cost of operation and absence of obnoxious fumes. The Moebius and Balbach processes are essentially silver refining processes and find application in the treatment of silver containing small amounts of gold. Full details of these methods will be found on page 288, under silver refining processes.

Another process arose out of the work of Dr. Emil Wohlwill in 1874 at the Norddeutsche Affinerie in Hamburg. At one time it occupied a very important place in gold refining, the bulk of the world's gold being treated by its means. The chief defect of the process is the lengthy time of treatment, leading to a high lock-up of gold and a resulting loss of interest charges. Largely for this reason it was superseded by the Miller Chlorine process at the turn of the century.

In the Wohlwill process gold bullion is cast into anodes measuring approximately $6\frac{1}{2}$ in. long, 3–4 in. wide, and $\frac{1}{4}$–$\frac{1}{2}$ in. thick, weighing about 90–110 oz. These are then suspended in porcelain cells from horizontal supporting rods. Cathodes are of pure gold sheet, and the electrolyte is a solution of gold chloride ($AuCl_3$) containing 50–60 gm per litre of gold and with 5–7% of free hydrochloric acid maintained at a temperature of 60–70° C.

The current used is 150 amperes at 12 volts, the current density being of the order of 100 amperes per sq. ft. Under the action of the current the gold together with base metals and any platinum pass into solution, the silver, however, reacting to silver chloride which falls to the bottom of the cell. If the anodes contain more than 6% of silver a coating of silver chloride is formed which renders them practically insoluble and inhibits further action. The application, however, of a pulsating current makes it possible to refine bullion containing up to 16% Ag, the silver chloride formed at the anode flaking off and leaving the surface of the gold free for further solution. A pulsating current is obtained by superimposing alternating currents on direct current by connecting a d.c. dynamo in series with an a.c. dynamo.

An anode weighing 90 oz. is dissolved in about two days, when the stump is removed and replaced by a new one. Cathodes are removed when they are about 80 oz. in weight, washed free from gold chloride, dried, melted and cast into bars assaying 999·5 fine.

The silver chloride anode slime is reduced with iron in hydrochloric acid solution and together with the anode residue is cast into fresh anodes. This material may contain up to 25% gold and is electrolysed in Moebius or Balbach cells.

Any platinum present dissolves from the anode with the gold, but is not deposited on the cathode. It is precipitated in the solution from time to time by the addition of ammonium chloride which precipitates the platinum as the yellow salt ammonium chloroplatinate which is filtered off, dried and reduced to metallic platinum on ignition.

SILVER

Like gold, silver was in use, and even current as money, in the remotest time.* Doubtless it was mined as it is today in conjunction with other metals, but until modern metallurgical

* Abraham paid Ephraim in silver for the land which he bought for the burial-place of his family. Genesis xxiii. 15 and 16.

methods made possible the separation of the constituent metals in complex ore deposits, most of the silver used in early civilization must have been mined in the native condition as at Laurium, near Athens, in 483 B.C.

Nowadays, most of the world's silver is produced as a by-product in the refining of other metals. It occurs associated with gold, copper and lead, and is recovered in the latter stages of refining these metals.

Many are the processes which have been applied to effect recovery of silver from its ores, amongst these being amalgamation, fabulous amounts of the metal being treated by this method during the occupation of South America and Mexico (1520–1820) by the Spaniards. Of these amalgamation processes, the Patio was perhaps one of the most widely used. It was introduced into Mexico in 1557 by Bartolome-Medina, a miner of Pachuca and was employed right up to the end of the 19th century, many millions of ounces of silver being recovered by its means. The method was especially suitable for dry arid districts where both water and fuel were scarce, and was applicable to siliceous ores consisting of native silver, silver chloride or argentite (Ag_2S). Extraneous metals such as pyrite, and galena do not interfere.

The ore was crushed and ground in a stone grinding-pan known as an arrastra (from the Spanish verb, *arrastrar*—to drag) which is essentially a circular pavement (Fig. 21) of stone surrounded by a low stone wall in which revolves a vertical shaft carrying four arms. At right angles to each arm are attached large stones weighing from 10–12 cwt., which act as mullers. When the shaft revolves, the stones are dragged round and exert a grinding action. Motion was imparted by mules, water wheels, or when available, by steam power. When the ore had been ground to a fine mesh it was run out on to a stone floor (*patio*) and allowed to dry to the consistency of clay.

It was then made into heaps or *tortas* about 12–24 in. thick and containing 100–200 tons of ore, 5–6% of salt being spread over the surface and incorporated with the torta by driving mules over the heap. On the following day 3% magistral (roasted copper pyrite, i.e. copper sulphate) was added and incorporated as before, following which 6–8 lb. of mercury for each pound of silver in the ore was added and mixed in, the heap being trodden by mules or turned over with shovels until amalgamation was complete; this took up to three weeks, the appearance of the amalgam indicating to an experienced operator the degree to which amalgamation had taken place. The amalgam was then separated from the gangue material by washing with water in a tank, the heavy amalgam settling to the bottom, the slime overflowing to waste. The amalgam was then collected, washed, the excess of mercury removed by squeezing through a chamois cloth, and then retorted, mercury distilling leaving behind the silver which was cast into bars.

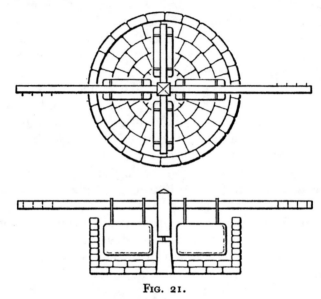

Fig. 21.

ARRASTRA ORIGINALLY USED IN MEXICO FOR CRUSHING SILVER ORE

The reactions are somewhat complicated in character and incompletely understood, but the following equations represent approximately the changes which took place. The first reaction is that of salt on copper sulphate

$$2NaCl + CuSO_4 = CuCl_2 + Na_2SO_4$$

the cupric chloride then reacting with cupric sulphide

$$CuS + CuCl_2 = Cu_2Cl_2 + S$$

the cuprous chloride—which is the chief reducing agent in the process—then reacting with the silver sulphide

$$Cu_2Cl_2 + Ag_2S = 2Ag + CuCl_2 + CuS$$

284

Silver chloride reacts with mercury

$$2AgCl + 2Hg = 2Ag + Hg_2Cl_2$$

Mercury thus transferred into calomel (Hg_2Cl_2) is the chief item of loss of mercury, which averaged about 1–2 oz. for each ounce of silver present in the ore.

A variation of the Patio process was invented by Alvaro Alonzo Barba in 1609. Barba, a Spanish priest, was curate of the parish of San Bernando in Potosi situated in the heart of the rich silver-mining district. His process was still in use in Bolivia, Peru and Mexico in the latter part of the 19th century. The process takes its name from the copper vessel (*cazo*—copper pan) in which the process is conducted, copper being the active chemical reagent concerned. Finely ground ore, water and salt were charged to the pan and then heated to the boiling-point of the pulp. Mercury was then added and the pulp stirred continuously. After several hours the pan was removed from the fire, the slime washed away with water and the amalgam recovered and retorted.

The silver chloride present in the ore was dissolved by the hot brine and reduced to the metallic state by the copper vessel and then amalgamated with the mercury. Any silver sulphide minerals were reduced as in the Patio process by cuprous chloride. The advantages of the process (compared with the Patio) were the very much shorter time (10–20 hours) and much lower consumption of mercury.

A modification of the Cazo process known as the Washoe or pan process was introduced about 1860 at the Comstock mines in Nevada. It employed a cast-iron pan of capacity about 1 ton of ore provided with mullers for grinding and mixing and steam jackets for raising the temperature of the pulp, thereby assisting amalgamation. The process took its name from the district in Nevada in which it was first introduced.

The crushed ore together with sufficient water to form a thick pulp was charged to the pan and the mullers set in motion. The ore was ground for about an hour or so and mercury then added, the amount ranging from 150–350 lb. per charge of about a ton. When amalgamation was complete, the contents of the pan were flushed into a settling tank, the amalgam collected and retorted for the recovery of silver. The extraction with 'free milling' ores attained 85–90%, but with sulphide ores the extraction decreased to 65–70%.

Metallic iron derived from the wear of the cast-iron mullers and from the iron of the pan itself was the chief reactant:

$$2AgCl + Fe = 2Ag + FeCl_2$$

the metallic silver amalgamating with the mercury.

In a modification known as the Boss process the ground material from the stamps flowed through a series of amalgamating pans and settlers, the amalgam accumulating in a separator attached to the front of each pan and settler. This had the advantage of making the process continuous and hence saved time and labour in transferring charges from pans to settlers.

Silver from the famous Comstock lode in Nevada, discovered in the late 1850s, was originally worked by the Washoe process, a hilarious account of which appears in Mark Twain's book *Roughing It*.

Roasting of Silver Ores. Complex ores of silver such as the sulphide, sulpho-arsenides and sulpho-antimonides cannot be successfully amalgamated without some form of previous treatment. This treatment commonly consists of roasting the crushed ore with about 10% salt whereby the silver is converted into chloride

$$2NaCl + Ag_2S + 2O_2 = 2AgCl + Na_2SO_4$$

In chloridizing roasting the best results are obtained when the ore contains from 3–8% sulphur. If too much is present the excess must be removed by a preliminary roasting without salt, otherwise a considerable amount of base metal chlorides will be formed if too little is present as in the case of oxidized ores, sulphur is usually added in the form of pyrite. As extensive mechanical equipment, fuel and skilled labour were essential for executing the operation it is not surprising to learn that chloridizing roasting of silver and other ores was initiated in Europe about 1785 and used on a large scale near Freiberg in Germany in 1790. At Freiberg the chloridized ore was rotated in a barrel together with water and scrap iron, the silver being reduced to the metallic state. After two hours or so of rotation, the barrel was opened and mercury added and the rotation continued for some 20 hours. The contents of the barrel were then washed out and the amalgam collected. Although the extraction was excellent, the silver obtained was of very low grade due to contamination with base metals which were also chloridized during roasting and reduced by the scrap iron. Attempts to volatilize the base metal chlorides by roasting at a high temperature resulted in an increased loss of silver, whereas slow roasting at a low temperature led to poor chloridizing of silver ore. These and other snags finally led to the abandonment of the barrel process and its substitution by leaching processes.

LIXIVIATION PROCESSES

The pan amalgamation processes were well adapted to small working with rich ore, but its high unit cost made it unsuitable for large-scale working and led to its gradual supersedence by leaching processes. These included the Augustin, Von Patera and Ziervogel processes which are now all obsolete, having been replaced by the cyanide process.

CYANIDATION

A gradual disuse of the above processes was brought about by the depletion of surface and suitable ores and high operating cost. Low-grade ores could not be treated economically at a profit by these methods. With the introduction of the cyanide process towards the close of the century the way was opened for the treatment of these low-grade ores on a large scale at low cost.

In Mexico especially, where low-grade silver ores were abundant, there was a promising field for the adoption of the cyanide process and it was in this country that the real pioneer work in the adoption of the process was carried out. One of the pioneers, E. M. Hamilton, in 1900 built and operated a large scale experimental plant for the treatment of the silver sulphide ores of the Sirena mine at Guamajuato, Mexico.

At a few of the older mills in Mexico and elsewhere, the ore is still crushed by stamps equipped with 30–40 mesh screens, and the pulp separated into sand and slime by cone classifiers. The sand is charged to tanks and leached for about 14 days with 0·4% KCN, the slime being thickened, agitated with cyanide solution, filtered and the solution together with that from the leaching tanks passed through zinc boxes or treated with zinc dust, the silver precipitate then being run down with a borax flux to bullion. In the more modern plants the recovery of silver by cyanide is similar in its application to that of the gold all-sliming process as described on page 277.

Argentiferous Base Metal Ores. A majority (75%) of the world's production of silver is not obtained from the cyanidation of silver ores, but from lead, copper, and other ores with which it is associated as a fortuitous constituent. The main source of by-product silver results from the refining treatment of argentiferous lead bullion which may contain up to as much as 300 oz. of silver (and 1–2 oz. of gold) per ton most of which is removed as a silver-rich lead alloy, cupellation resulting in a 990 fine product.

In addition practically all the base metals which are refined by electrolysis, e.g. copper, nickel, etc., contain silver and other precious metals which precipitate as a slime in the cell bottom.

In the case of electrolytic slime, preliminary operations are necessary in order to effect the collection of the precious metals in a Doré, the steps being designed to remove as much as possible of the base metals. The steps involve:

1. Roasting to oxidize the base metals.

2. Water or acid leaching of the calcine to remove soluble metals.

3. Smelting of the insoluble lead residue with flux to slag off the remaining base metals.

4. Doré metal from (3) is cast into anodes which form the raw material for electrolytic parting.

In early days nitric and sulphuric acid were used but these have now given way to electrolytic processes employing a silver nitrate electrolyte.

A process for the electrolytic parting of silver and gold was patented by Moebius and introduced into Mexico in 1884 and later in many silver refining works in America and Germany.

The Doré metal is cast into anodes which measure 10–15 in. long, 6–10 in. wide and $\frac{1}{2}$ in. thick and weigh 150–200 oz. and suspended vertically inside the cell by hooks from a conductor bar. Each four anodes are enclosed in linen or cotton bags stretched over a wooden frame for the purpose of catching the gold and insoluble impurities detached from the anodes, and of preventing contamination of the deposited silver. Cathodes are of thin rolled plates of pure silver or of stainless steel. The cells are of earthenware or plastic material, a removable wooden tray being placed at the bottom of the cell to receive the silver dislodged from the cathodes by mechanically operated wooden scrapers. The reason for the provision of this scraping gear is that as the distance between anodes and cathodes is small, the deposited silver if left unchecked would grow outwards from the cathodes to the anodes and eventually cause short-circuiting.

Electrolyte is a weak solution of silver nitrate, containing 60–80 gm per litre, aciduated with about the same quantity of nitric acid. Voltage of the cells is 1·5 and a current of 300 amps is used corresponding to a current density of 30–40 amps/sq. ft.

The silver recovered per cell is about 800 oz. per day assaying 999·5 fine. The slime in the bags consisting of gold, silver, lead and other insoluble metals is removed once or twice a week, and treated with sulphuric acid to dissolve out the silver leaving a residue from which the gold may be recovered electrolytically or by acid parting. Silver in the solution obtained by digesting with sulphuric acid is recovered by cementing out on copper sheets immersed in the solution.

In the Balbach-Thum process, initiated by E. Balbach (1839–1910), the electrolyte is the same as that used in the Moebius but the anodes instead of being suspended vertically in the cells are placed horizontally in wooden trays supported on the edge of the cells. The trays are lined with canvas to retain the slimes. Cathodes are of graphite in sheet form $\frac{1}{2}$ in. thick and form the bottom of the cells.

In general the Balbach process is employed when appreciable amounts of gold and platinum are present, necessitating frequent removal of the slimes. When only small quantities of gold are present, the Moebius is more economical. A comparison of the two processes is given by A. E. Richards* as follows:

(a) The Moebius process leaves anode stumps which have to be remelted. The Balbach cell consumes all the anode.

* 'Refining of Gold and Silver'—*Refining of the Non-Ferrous Metals*, p. 111, I.M.M., London, 1950.

(*b*) The Moebius cell cannot be conveniently used for low-grade anodes, whereas the Balbach cell can.

(*c*) The voltage of the Moebius cell, and therefore the power cost, is lower than the Balbach.

(*d*) The output of the Moebius cell per sq. ft. of floor space is higher than the Balbach.

(*e*) The Balbach cell has no moving parts, is easier to maintain and requires less labour.

(*f*) the risk of contamination of the cathode deposit with anode slimes and consequent loss of gold is much greater in the Moebius cell than in the Balbach because there is greater danger of split anode bags.

(*g*) The Balbach cell is slow and involves a greater lock-up of silver.

THE PLATINUM METALS

The six metals of the platinum group form a well-defined series of elements usually found associated in the metallic state. The members fall readily into two groups of three each: platinum, iridium and osmium with high atomic weights and high densities ranging from 22·4 to 21·4; and palladium, rhodium and ruthenium with densities ranging from 12·4 to 11·4. Platinum and palladium are the two principal members of the group and occur more abundantly.

Native platinum appears to have been first discovered during the 16th century in placer deposits in the Choco district of Colombia, South America, then part of the Spanish colonial empire. The platinum was found associated with gold and was regarded by the Spaniards as a nuisance for it demanded a great deal of effort to separate it from the gold. When present in quantity it rendered the recovery of gold uneconomic and the workings were abandoned.

The metal received its name in consequence of its white appearance and close resemblance to silver, the Spaniards calling it *platina del Pinto*, *platina* being the diminutive of *plata*, the Spanish for silver, Pinto being the name of the river in the auriferous sands of which the metal was first discovered. Specimens of the metal reached Europe early in the 18th century.

The fact that native platinum was a mixture of metals was not immediately recognized, its complexity causing much difficulty in its use. Subsequent investigations, however, revealed the true nature of native platinum, namely, that it consisted of a complex of five other metals. W. H. Wollaston (1766–1828), a doctor of medicine, first isolated palladium in 1802 and rhodium two years later, named the first after Pallas, a planet that was discovered about the same time, and the second from the Greek *rhodon* meaning a rose because solutions of rhodium salts are rose red in colour. Osmium and iridium were both discovered in 1803 by S. Tennant (1761–1815), professor of chemistry at Cambridge, a contemporary and friend of Wollaston. Osmium was named from the Greek *osme* meaning smell on account of the penetrating odour of the volatile oxide, iridium taking its name from the Greek *iris*, a rainbow because of the differing colours displayed by its salts. Ruthenium is the youngest of the group from the viewpoint of its date of discovery. It was first isolated in 1844 from platinum ores obtained from the Urals and derived its name from Ruthenia, the Latin name for Russia—the adopted country of Carl Claus, its discoverer.

Occurrence. The chief platinum minerals are native platinum, osmiridium or iridosmine and spherrylite, an arsenide ($PtAs_2$) discovered in 1875 by Sperry in an outcrop of ore at Sudbury. Numerous assays show that as a rule the percentage for the different constituents in native platinum is approximately as follows: Pt 65–90%, Pa 0·5–3%, Os up to 5%, Ir up to 5%, Rh up to 1%, Ru 0·5%. Of the base metals Cu occurs up to 4%, Fe 2–10%; small amounts of Ni and Co may also be present.

The composition of osmiridium likewise varies considerably, its constituents being osmium 26–30%, iridium 25–30%, ruthenium 10–15%, platinum 10–16%, rhodium 13–16% and gold 0·1–0·7%.

The platinum metals occur in both lode and alluvial deposits in the following associations:

(*a*) Igneous rocks as in the Urals and the Transvaal which are worked essentially for their platinum metals.

(*b*) Sulphides of Ni, Cu and Fe, notably nickelferous pyrrhotite, as at Sudbury, Ontario, and in the Transvaal. At Sudbury the platinum metals are obtained as by-products during the extraction of nickel and copper.

(*c*) Placer deposits as exemplified in the Perm district of the U.S.S.R. and in Colombia, Alaska and Brazil.

Up to the end of the 18th century the placer deposits of Colombia constituted the chief source of the metal. In 1824, native platinum was found in the gold placers of the Goroblagodat and Nizhny-Tagil districts in the Urals. These latter deposits contained up to 85% platinum accompanied by gold and silver, and Russia became the chief producer of alluvial placer platinum, a position which as far as is known she still holds. Platinum from the Sudbury deposits first appeared in the market about 1910, production increasing until by the mid-1930s, Canada became a dominant factor in world supplies. In the early 1920s platinum was discovered in the Rustenburg district in the Transvaal, production having gradually increased until now South Africa forms one of the world's largest producers.

The progressive rise in world output can be seen from the following table.

WORLD OUTPUT OF PLATINUM IN TROY OUNCES

Country	1915	1926	1960
Canada	475	9,521	460,000
Colombia	18,749	55,000	28,850
U.S.S.R.	104,000	92,700	275,000 (Estimated)
South Africa	—	4,951	400,000
Others	96,709	5,328	26,150
	219,933	167,500	1,190,000

RECOVERY

Concentration of platinum in placer deposits takes advantage of the high specific gravity of the metal, early separation being effected by washing with water in sluices, rockers, etc., the gold which concentrated with the platinum being removed by amalgamation. Dredges were introduced into Russia about 1896 and in Colombia about 1915. Gravity concentration using shaking tables is practised in South Africa on osmiridium and platinum ores, subsequent flotation separating the base metals.

On the Sudbury field in Canada the nickel ore, after concentration and separation of copper by flotation, is roasted, smelted and converted to matte and refined by either the electrolytic or the Mond carbonyl process. Platinum and precious metals are separated from the main nickel and copper refining processes at four points, (1) the undissolved slimes remaining in the electrolytic tanks used for refining of nickel, (2) similarly in the electrolytic slimes from copper refining, (3) as a precious metals alloy separated from the ground matte by magnetic means, (4) in the final residues from the volatilizers in the Mond process operated at the Clydach refinery in Wales. In all of these residues the content of platinum metals is small and accordingly they are given a preliminary treatment for removal of the bulk of the other metals before the actual extraction of precious metals takes place.

In South Africa the platinum metals flotation concentrate is fed to a blast furnace for smelting with coke and flux to a matte, which on conversion yields a high grade copper nickel matte, subsequent electrolysis yielding a precious metals sludge which is collected and dried.

Refining. The refining process is based on the wet methods used by Wollaston and Tennant in 1802–04. The method consists in treating the concentrate with mineral acids to dissolve the various metals selectively. Depending on the nature and amount of precious metals present, variations of the method are possible. All are somewhat complicated.

The main version depends on the solubility of platinum, palladium and gold in aqua regia ($HCl + HNO_3$), whereas rhodium, ruthenium and iridium remain unattacked. The platinum concentrate is dissolved in aqua regia, the insoluble residue separated from the solution of the chlorides from which gold is precipitated by ferrous sulphate as a brown powder which is washed free from iron and then melted and cast into bars. Addition of ammonium chloride to the solution precipitates the platinum as yellow ammonium chloroplatinate, further addition of the ammonium salt followed by excess hydrochloric acid resulting in the precipitation of palladium. On ignition at a low temperature both the platinum and palladium are converted into small silvery-grey nodules with a texture resembling that of coke, material in this condition being known as sponge metal. On subjecting to high pressure and sintering, compact metal results assaying 99·9%. The process for the separation of the other metals of the group is as follows:

(1) The insoluble residue remaining from the treatment with aqua regia is fused with salt in a current of chlorine, dissolved in water, acidified with hydrochloric acid, and hydrogen sulphide passed in. Rhodium and ruthenium precipitate, iridium remaining in solution.

(2) After filtration the filtrate is made alkaline with caustic soda and chlorine passed in. Iridium oxide precipitates which is collected, dried and reduced to metal by hydrogen.

(3) The ruthenium and rhodium sulphides from (1) are treated with hydrochloric acid and chlorine, with alkali added, rhodium separating as the hydroxide which is reduced to metal by reduction with hydrogen.

(4) Solution containing the ruthenium is treated with alcohol which causes it to precipitate as a black flaky oxide which is reduced to metal as in (3).

(5) Any osmium remaining in the residue from the initial treatment is recovered by subjecting to a temperature of 600–700° C. in a current of hydrogen. Volatile osmic tetroxide is formed which is absorbed in a solution of caustic potash and the osmiate precipitated by addition of alcohol.

The insoluble residue obtained from the preliminary aqua regia treatment can also be treated by fusion with sodium peroxide or a mixture of barium peroxide and barium nitrate. Both these methods oxidize the metals yielding compounds soluble in acids.

If relatively large amounts of gold and silver are present, as in the Clydach residues, a preliminary smelting is given in which lead acts as a collector of the precious metals. Subsequent cupellation results in an alloy rich in platinum, gold and silver. This is parted with sulphuric acid which removes most of the silver and about one-third of the platinum. The remainder of the precious metals are then recovered from the residue by the aqua regia treatment.

Usage. Originally the main demand for platinum came from jewellers for ornamental purposes, in fact its first use for trinkets, spangles, beads, etc., originated in the 16th century. For centuries, however, very little interest was taken in the metal because its high melting-point made it difficult to work. Eventually, however, it was realized that the very properties that made the metal difficult to produce—high melting-point (1773° C.) and resistance to the action of almost all single acids —fitted it well for a number of industrial uses in addition to jewellery, and in fact this latter use is now only a minor outlet.

The earliest (1830) major employment of platinum was for corrosion-resistant evaporating pans for concentrating sulphuric acid, this operation originally being carried out in small glass vessels.

Modern uses of the metal stem from its high resistance to chemical corrosion especially at high temperatures, good electrical conductivity and catalytic properties. The electrical industry uses platinum for switchgear contacts, the absence of surface film formation ensuring low contact resistance. In the chemical industry advantage is taken of its resistance to corrosion by using it to make or to line crucibles, dishes, capsules, tongs, etc.

Platinum and its alloys act as catalysts in many chemical reactions, the most important being the conversion of sulphur dioxide to sulphur trioxide in the manufacture of sulphuric acid and the oxidation of ammonia to produce nitric acid. In the petroleum industry a platinum catalyst is used to increase the octane value of petrol. For catalysis a large surface is required so that the metal or alloy is usually spread over the surface of some other material such as silica gel, or used in the form of gauze. Palladium is used in jewellery, the so-called white gold being an alloy of gold decolourized by the addition of palladium.

The percentage usages are approximately as follows:

Chemical industry	51%
Electric and electronic	33%
Jewellery	10%
Dental	5%

The minor platinum group metals, iridium, rhodium, osmium and ruthenium are used principally as alloying elements to improve the properties of platinum and palladium, small additions increasing the hardness, tensile strength and resistance to heat and corrosion. Rhodium in addition to being used to form a hard alloy with platinum is used as a coating material; the hardness and lustre of its surface makes it ideal for plating such metals as silver to which it gives an untarnishable finish.

Ruthenium is mostly used as a hardener for palladium and platinum. It is contained in the hard alloys of osmium used for tipping the gold nibs of fountain-pens.

BIBLIOGRAPHY

1. CLENNELL, J. E., *Cyanide Handbook*, 1910. McGraw-Hill Book Co., New York.
2. HAMILTON, E. M., *Manual of Cyanidation*, 1920. McGraw-Hill Book Co., New York.
3. RICHARDS, A. E., *Refining of Gold and Silver: The Refining of the Non-Ferrous Metals*, 1950. Institution of Mining and Metallurgy, London.
4. PRENTICE, T. K., *Metallurgical Practice on the Rand*. Bulletin No. 367, April 1935. I.M.M., London.
5. COLLINS, H. F., *Metallurgy of Lead and Silver*, 1900. C. Griffin & Co. Ltd., London.
6. PERCY, J., *Silver and Gold*, 1880. John Murray, London.
7. McDONALD, D., *A History of Platinum*, 1960. Johnson Matthey & Co. Ltd., London.
8. *Platinum and Allied Metals*, 1936. Mineral Resources Dept. Imperial Institute, London.

CHAPTER EIGHT

SHAPING OF METALS

EXTRACTIVE metallurgy finishes with the production of
metal in a pure or approximately pure state. Subsequent
operations have to be performed on the metal to shape it
into a suitable form for the consumer. The metal may be
remelted and cast into the form in which it will be used,
alternatively it may be cast into a simple shape such as an ingot,
slab, or wirebar for subsequent working by the application of
mechanical force. Mechanical shaping of ingots may be performed by hot forging, hot or cold rolling or cold working into
wire, pressings, etc. The greatest tonnage of worked metal is
produced by rolling, but forging under the hammer or hydraulic
press is important, and extrusion is also used extensively in the
shaping of metals. Plates, sheets, strips, tubes, wire, bars and
rails can all be produced by mechanical shaping. In a few
instances the worked products such as tubes and rails are in the
form in which they are used; other products such as sheet, strip,
etc., form the raw material for engineering industries and
require further processing.

SHAPING BY CASTING

Giving shape to an object by pouring it in liquid form into a
mould is one of the oldest of the metallurgical arts, for it was
practised by the Egyptians of the first dynasty as far back as
5000 B.C. A high degree of skill was attained as many fine
examples in museums bear witness.

Earliest castings were made in flat open stone moulds of
simple shape following the implements in use at the time such
as spear heads, axes and knives. Permanent metal moulds
were introduced some time in the Middle Bronze Age;
the finest bronze castings, however, including some of the
most splendid statues ever made were produced by the *cire*

perdue or wax casting method now known as investment casting.

Compared to bronze castings those of cast iron are of comparatively recent date, for primitive communities were not able to heat iron to a sufficiently high temperature for it to become molten. Sand introduced in the 18th century was found a much better moulding material than the clay previously used and from those times until the present there has been an extension of the range of moulding material from clay with small amounts of sand to sand bonded with a small amount of clay to make it plastic. To contain the sand, moulding boxes, or flasks as they are now called, were constructed consisting of a frame of iron or steel rectangular or circular in outline, having a top half or cope (cap) and a bottom half or drag, fitted together by means of two or more pins.

The hole in the cope through which the metal is poured into the mould is known as the runner or downgate. Risers are replicas of the downgate, the metal running through the mould, carrying with it any debris or dirt overflowing out into the riser. Risers also act as feed gates, i.e. they feed the casting as the metal contracts by allowing it to flow back into the mould, thus filling the shrinkage cavities.

Hollow castings are made by making a core of sand bonded together by a suitable material such as molasses or dextrine and then baking. The hard core is then supported in the mould by appropriate means. On pouring, the metal fills the space between the core and the inside of the mould.

Up to the beginning of this century, moulds were made in this time-honoured manner entirely by hand. In recent years, however, the demand has necessitated the substitution of mechanized methods for hand-moulding, many ingenious mechanical devices having been developed to facilitate the speed and precision with which a moulding can be made. Thus for the rapid filling and simultaneous ramming of large moulds, a machine known as the sand slinger has been developed which rapidly impels sand into the mould at a rate of up to 20 cu. ft. per minute. Jolt ramming machines compress the sand in a manner similar to hand-ramming. By these and other devices complete moulds can be produced at the rate of up to 60 per hour and in many foundries mechanization has been achieved in all stages including moulding, core-making, pouring, knocking out and sand reclamation.

Cast Iron. British iron foundries produce about 4 million tons of castings per annum, which represent many times the weight of all non-ferrous and steel castings combined.

The demand for cast iron arises from its cheapness as an engineering material combined with the properties of wear and abrasion resistance and machinability. In addition, the high fluidity of cast iron simplifies the casting operation. Early castings were usually made from metal direct from the blast furnace, the molten pig iron being run into shaped open sand moulds.

Pig iron, however, is not a material of constant composition, for depending upon the mode of production, the type of fuel used in the furnace, the nature of the ore, etc., the chemical composition of the iron will vary and little control could therefore be exercised over the quality of the cast iron. With the invention of the cupola by W. Wilkinson, a Shropshire iron-master, in 1795, the situation changed and the separation of smelting from founding was made possible. By remelting pig iron a cast iron of exact composition could be obtained by appropriate blending of pig iron and other material. Special variations of cast iron could also be produced by selection of different pig irons, by special alloying addition and by variation of melting conditions within the cupola.

The cupola in effect resembles a small blast furnace and consists of refractory-lined vertical steel shaft 15–20 ft. high, 4–6 ft. in diameter. Air is supplied from a fan or blower via tuyères situated at a height of 2 ft. or so from the bottom. Charging is done through an opening at a height of 10–15 ft. above the tuyères. Metal is tapped from the furnace at the bottom. Coke is charged to the level of the tuyères and set alight. Alternate layers of metal and coke are then charged, limestone as flux being charged with the coke, the amount used varying from 20 to 30% of the weight of the coke. Heat necessary for the reaction is supplied by combustion of the coke with the air. Limestone combines with the sand and rust adhering to the pig and scrap iron and also with the coke ash, and forms slag which is periodically tapped from the slag notch, metal being tapped, when a sufficiency has accumulated in the hearth, to ladles whence it is conveyed to the casting bay.

As the metal in the cupola is in contact with coke the main tendency is to pick up carbon, and when steel scrap is included with pig iron the carbon content of the cast metal is higher than that of the charge. The amount thus absorbed is, however, restricted to the solubility of carbon in iron which is about 3·5%. With no steel scrap, however, the carbon content depends more on the operating conditions and the presence of other elements.

In recent years there have been several developments in cupola practice aimed at reducing labour and material costs and also at meeting increasing competition from the electric melting furnace. Water cooling in the melting zone has been introduced thus effecting a saving in maintenance and refractories. Economy in coke consumption is being increasingly achieved by use of hot blasting, the combustion air being preheated by utilization of the sensible heat of the waste gases. Replacement of hand-charging by automatic charging employing skip hoisting is also a noticeable feature.

Gravity Die Casting. In gravity die casting, molten metal is poured into a metal mould normally made of iron or steel. The metal moulds called dies are used over and over again. The process of gravity die casting has advantages which follow from the permanent nature of the mould. A sand mould is used for one cast only and is destroyed in extracting the casting; the average life of a die on the other hand may be something of the order of 20,000 casts. Dies are clearly much more expensive to make than patterns for use with sand moulds, hence their use is only justified when castings are required in large numbers. Production is much more rapid than with sand moulds, which leads to considerable economies. One of the earliest applications of the permanent mould was the production of toy soldiers which first appeared in the 1800s and were made by gravity pouring of molten lead into iron moulds constructed in two halves, which were hinged together and mechanically opened and closed.

Pressure Die Casting. In this type of casting molten metal is forced into a metal mould under pressure. It is the fastest of all casting processes and is employed when rapid production of small accurately dimensioned components is necessary, for use in motor cars, refrigerators, washing machines, etc.

It may be said to have originated in connection with the casting of printing type about a century ago and eventually led to the development of the linotype machine by O. Mergenthaler.

In 1849 Sturgiss patented a manually operated machine in which molten metal was forced into a die by a vertical ram actuated by a hand-lever. Its first use was for type casting. Thereafter the development of die casting machines followed along two different channels, differentiated by the kind of pressure employed in forcing the metal into the die. In the plunger or ram type as exemplified by the Sturgiss machine the metal is forced into the die by means of a ram immersed in the metal, whereas in the pneumatic type air pressure is employed.

Early die casting was carried out principally with lead or tin and alloys based on them; these appear to be the only metals employed for die casting up to the beginning of this century, the reason being that they cast readily at relatively low temperatures. Although possessing good bearing metal properties and excellent corrosion resistance, the tin and lead alloys are soft and low in strength and the former is very expensive which accounts for the fact that today these alloys, except for type metal, are no longer pressure die cast. The decline in the importance of lead and tin alloys for die casting began about 1916 with the advent of the harder, stronger zinc alloys; in fact pressure die casting since that date has been closely associated with the development of suitable zinc alloys. Application of aluminium alloys for commercial die casting dates from about the same period, but the higher melting-point of aluminium (659° C.) and its alloys introduced new problems. Thus the tendency of aluminium to dissolve iron made it impossible to use a plunger for injecting the metal into the die. This led to the development of gooseneck machines depending upon direct air pressure rather than a ram for metal injection, the gooseneck chamber serving to contain the liquid metal temporarily until forced into the mould.

Successful pressure die casting of brass and other copper base alloys was developed in Europe about 1930. The high melting-point of copper (1083° C.) makes these alloys difficult to die cast. Gooseneck machines are not suitable nor is the plunger type commonly used for zinc alloys, both types of machines having the disadvantage that they produce porous castings containing iron to a variable degree. Iron is an undesirable impurity, promoting brittleness and reducing the ability of castings to withstand the corrosive effects of various chemicals in service.

The equipment developed to deal with this nuisance was the so-called 'cold chamber machine'. The chamber is a cylinder fitted with a ram operated by hydraulic pressure, the metal being ladled into the chamber whence it is forced into the die cavity by the advancing plunger. The injected alloy solidifies in the die, half of the die is withdrawn, the casting being pushed out by ejector fins, the two halves of the die being brought together again, and the operation recommences. High temperature results in relatively short life of the die; hence production of pressure die-cast brass, etc., is limited, as compared to that of zinc and aluminium. Since 1935 the cold chamber machine has been applied to aluminium alloys and in fact has largely supplanted gooseneck machines, chiefly because of denser castings which do not suffer from the defect of iron contamination.

Some 75% of total pressure die casting is in zinc alloys. Of the remainder about 75% is in aluminium alloys, the residue being tin, lead and magnesium alloys. Machines in common use range in pressure from 3,000 to 15,000 lb. sq. in. up to 30 castings per minute being produced.

In general the method is limited to small articles but recent developments include a machine that will accommodate a casting area of up to 600 sq. in. giving a maximum weight of 30 lb.

INVESTMENT CASTING

Although sand casting is capable of producing castings to most of the forms required in engineering practice, the complexity of certain shapes makes it impossible to remove the pattern. This is particularly true of statuary and other artistic or decorative objects. For such work the founder may have to turn to the ancient *cire perdue* or 'lost wax' process now termed 'investment' casting. In this process a wax model of the required shape is 'invested' by surrounding it with a plaster-like composition which is then allowed to harden. The hardened investment is then heated so that the enclosed model melts, the molten wax running out through prepared channels. Baking is then applied to give the investment the necessary strength for the next stage in which molten metal is poured into the cavity left by the outflow of wax. The metal then solidifies leaving the casting surrounded by the investment which is broken to release it. By this ingenious process the necessity for a two-part mould and loose cores is avoided. The method has an extremely ancient history being applied in Britain over 2,500 years ago by metal workers of the Bronze Age. Even earlier the method was well understood by the Chinese and Egyptians who used it for making filigree and other fine jewellery. The detail was so good that it was only when replicas of the same object were discovered that it was realized that repetition as opposed to hand-wrought articles was involved.

As a production technique, however, the lost wax process was virtually unknown to industry until the outbreak of World War Two. It was then discovered that the process was applicable to the mass production of jet engine turbine blades to the necessary very close dimensional tolerances. Previously the method of forming high-temperature turbine blading was by precision forging, but as the high-temperature strength of the alloys increased, so hot working became more difficult until the stage was reached when new high-strength alloys could not be economically forged. The solution was found by investment casting. Today virtually every branch of engineering has taken advantage of this development, products being used at a rate of approximately 20 million per annum mainly to replace components previously designed for production by other metal-forming techniques.

The advantages of the method are (1) accuracy; castings can be produced to very fine limits, (2) production of complex and difficult shapes with high melting-point metals difficult or impossible of attainment by other forming processes.

In its modern application wax or polystyrene is injected in liquid form under pressure into an accurately machined die which is made in two halves. After cooling and removal of the die, a wax or polystyrene pattern of the requisite shape remains. This is then either sprayed or dip-coated in a suspension of finely ground refractory material, and then dried and fired at a temperature of 1000° C. to melt or burn out the wax or polystyrene, leaving a cavity which is an exact negative of the casting required. Liquid metal is then poured through the feed aperture into the mould. When the metal has solidified the refractory shell is broken away to expose the metal casting which is then dressed and polished. Approximately 95% of production is within the weight range of a fraction of an ounce to 5 lb., but investment castings exceeding 50 lb. are now being produced. Metals so cast include aluminium, copper and nickel base alloys, iron and steel.

FORGING

The modern forging industry was born in the village smithy; though the tools employed and the materials are now much larger the principles are still the same. Metal is first heated, transferred to an anvil and shaped by repeated blows from a hammer. The largest piece of metal worked by the blacksmith does not exceed 40 lb. or so, whereas a heavy forge often handles steel ingots of 30 to 40 tons, the ingot being shaped by repeated blows from a hammer activated by steam power; on still larger forgings a hydraulic press is employed. For mass production of small forgings the system of drop forging is used. The main purpose of the operation is to forge metal into some form which cannot readily be achieved by rolling or other shaping operation.

Steam Hammer. The first known power hammer called a tilt was built in England and consisted of a horizontal pivoted wooden beam or helve provided at one end with a heavy iron hammer or tup. The beam was raised by means of a steam-driven cam rotating beneath the beam, the tup descending by its own weight on the hot work piece resting on the anvil. The power for working tilt hammers originally came from water-wheels; one of the factors determining the early growth of Sheffield as a steelmaking centre was the location of many nearby streams for driving the water-wheels.

In 1838 John Nasmyth, a Scots engineer, conceived the idea of a direct-acting steam hammer, for which he received patent rights in 1848. The development of this hammer, in addition to delivering a series of heavy rapid blows, also provided for the first time the means of forging the large masses of metal then being made available by the Bessemer and open-hearth steel-making processes. Hammers of 25 tons (weight of tup) were in operation in Sheffield as early as 1865. The steam hammer consisted of a two-piece frame supporting a steam cylinder, to the piston of which was attached the tup. By admitting steam to the cylinder below the piston the tup is raised a distance equal to the length of the cylinder, the steam then being discharged allowing the tup to drop upon the work piece supported on the anvil. To withstand the violent impact of the tup the weight of the anvil is many times that of the tup and cushioned by timber foundations. The main power for forging was thus derived from the inertia of the falling tup, the unit being known as the single-acting hammer.

Later, by the simple expedient of admitting steam also to the top of the cylinder, the double-acting hammer was pioneered. The effect of this device was to assist the hammer in its descent thereby intensifying the blow delivered by the tup, the steam thus being employed both on the descending stroke and on lifting the tup.

Forging Press. The steam hammer remained the chief means of working wrought iron and steel down to the middle of the 19th century and was also used as an alternative to primary rolling for large items such as plates, and for the rough shaping or 'cogging' of steel ingots prior to rolling. Steam hammers are quick acting, delivering sharp rapid blows, the effect of which is localized to the exterior of the metal and does not penetrate through heavy sections. For this reason it is limited to ingots ranging from half a ton to 15 tons capacity.

Sir Joseph Whitworth experienced trouble with the early Bessemer steel owing to the presence of air bubbles in the ingot and devised a process for hydraulically compressing ingots while still internally fluid. His first patent was taken out in 1856 and large hydraulic presses capable of exerting a pressure up to 2,000 tons for forging guns and armour plate were being installed soon after.

The press consists of an upright four-columned framework supporting a hydraulic cylinder, the ram of which carries the upper forging block or tool. By addition of water under high pressure to the top of the cylinder the ram is forced down exerting a steady squeeze, which penetrates the material to be forged much more deeply than does the blow of the steam hammer. Modern presses weigh up to 2,000 tons and exert pressure of up to 15,000 tons, the actual pressure applied to the work piece being of the order of 12 tons/sq. in. Press forging on a large scale is mainly carried out on armour plate, guns, large crankshafts, and boiler drums, their weights varying from 60 to 200 tons.

Drop Forging or Stamping. Forging of the type described above is mainly confined to the working of heavy masses of metal and is only justified when large components are being made. For the production of large numbers of small forgings to one pattern, forging between specially prepared dies is employed. This process dates from a few years before the American Civil War when E. K. Root and C. E. Billings of the Colt Arms Co., in Connecticut, developed a process for producing metal dies of sufficient accuracy to permit the making of small and intricate forged shapes. The parts were made by forcing hot metal into impressions formed in solid blocks of hardened metal, the so-called forging dies. The dies are made in halves containing in each half the shape of the final product, one half being attached to the tup and the other to the anvil. After three or four blows from the former the work metal fills the impression and has been stamped to the shape of the die. The method was first used in the production of firearms. Drop forging is practised not with ingot metal but with billets or bars of sizes appropriate to the product; it is used mainly in connection with steel components, but also in forging aluminium alloys, copper and brass. Typical examples are connecting rods, crankshafts, gear axles for cars, turbine discs and blades for jet engines.

ROLLING

Rolling involves the passage of metal between two horizontal iron or steel rolls which turn at the same speed but in opposite directions. As a result the metal increases in length at the expense of a reduction in thickness, the width, however, remaining practically constant. The method is widely used in the manufacture of sheet, strip, plate, rails, light and heavy sections, beams, joists, etc., in all metals both ferrous and non-ferrous, especially so in steel; about 90% of all steel produced receives its preliminary shaping by rolling.

In some cases the metal ingots are rolled directly to the finished product, but usually the heated ingots are first broken down in cogging or slab mills to lighter products of simple square (blooms) or rectangular shape (slabs). When further reduction or shaping is given beyond the bloom or slab stage, the material is passed through several stands (mills), a set of rolls and the housings in which they are mounted being known as a stand. Typical arrangements of these stands are:

(*a*) A train of stands, arranged side by side, and coupled together so as to use only one motor to drive them.

(*b*) A number of stands arranged in tandem, one after the other, so that the piece passes continuously through them receiving one pass in each. This type of mill is especially suited for high output with long runs of each size of product. The piece is elongated as it passes through each stand, and therefore its speed of travel increases during its passage through the rolls. Consequently, the speed of the stands for continuous rolling increases successively from the first to the last stand, to match the elongation of the piece.

(*c*) Mills for a high output may consist of several groups, each of one or more stands. Thus the first group, concerned mainly with reducing the piece to a smaller cross-section, comprises the 'roughing' stands, and the last group, concerned with obtaining accurate size and shape, is made up of the 'finishing' stands. Between these there may be some intermediate stands. Billets, small sections and bars, rods, wide strips, narrow strips and sheets may all be produced in continuous mills with up to 20 stands of rolls.

In most trains consisting of several stands side by side the piece is run straight out on either side of the mill, just as in the cogging mill; but when the piece is small enough in section, the time for running out on the floor can be saved by turning the front end into the next stand so that the piece forms a number of loops, the method being known as 'looping'. It can be done either by hand or mechanically by semicircular guide channels known as repeaters.

The semi-finished and finished rolled products are distin-

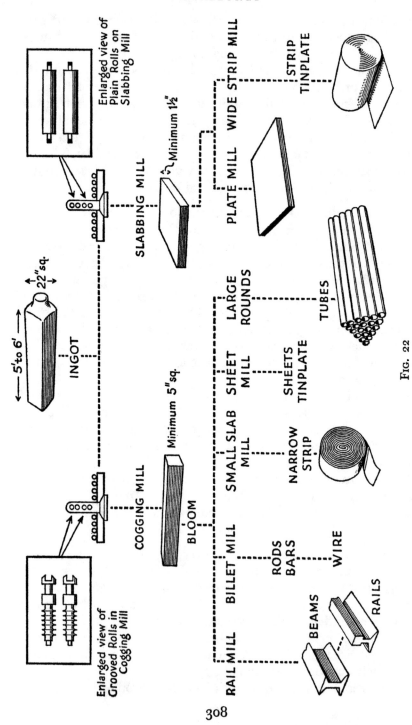

FIG. 22

PRODUCTS RESULTING FROM SEQUENCE ROLLING OF STEEL INGOT

guished by various terms, the more common being given in the diagram opposite.

The production of rolled metal dates back to small hand-driven mills of the 15th century, the method being restricted to metals soft enough to permit cold working, such as gold and lead. The earliest record of any mill for this purpose appears to be that in a sketch in one of Leonardo de Vinci's (1452–1519) notebooks. In the 16th century goldsmiths used a primitive form of mill to flatten pieces of metal for jewellery purposes and for the preparation of coinage blanks. One of the earliest applications of the rolling mill to the iron industry occurred in the 17th century, a mill being used for splitting iron into long narrow strips or bars for nail-making. Apparently this mill was developed in Sweden, being introduced into England in 1830 by R. Foley, a Stourbridge ironmaster.

By the end of the 17th century larger iron rolls were being cast enabling more powerful mills to be built, driven by water power. Further development was restricted by the power then available, and for the next 100 years or so there was little change in basic design, but with the advent of steam power towards the end of the 18th century further extension in both mill size and output took place.

In 1783 Henry Cort, the inventor of the puddling process for making wrought iron, introduced the grooved roll, as opposed to the plain roll, the openings in the roll thereby permitting the shaping of material into bars, rods and sections such as rails.

The method enormously increased production, output being 15–20 times that of the original hammering method. The first mill incorporating Cort's rolls was Crawshay's Works in South Wales, which by 1803 was producing 70 tons of bars per week. By closing the grooved rolls with large collars as in modern practice, J. Birkenshaw in 1820 patented the wrought iron edge rail which gradually superseded the cast iron type of rail.

Cort's mill was provided with a flywheel to boost the torque, the rolls turning in one direction only, the work piece having to be returned from the rear of the mill to the front side after each passage. This was done by a 'catcher' seizing the piece and lifting it bodily over the upper roll from where it was carried back to the original position and entered again in the next smaller groove of the rolls. James Nasmyth, of steam hammer fame, suggested that the flywheel be dispensed with and the rolls reversed after each pass, thereby eliminating the time-consuming operation and heavy manual work continuously involved in transferring the work piece from the rear to the front of the mill.

The idea was developed in 1866 by J. Ramsbottom, chief engineer of the London and North Western Railway Crewe Works, the reversal of the mill thus permitting the metal to be passed to and fro through the rolls.

In the United States the problem of passing the metal back, which British steelmakers had solved by the introduction of the reversing mill, had been solved in another way, by the invention in 1857 of a three-high mill by John Fritz at the Cambria Iron Co., Johnston, Pa. Fritz, by placing a third roll above the other two, made it possible to avoid this time-consuming operation, the metal passing in one direction between the top and middle rolls, coming back in the opposite direction between the middle and bottom rolls. As a result of the speed attained by this device the elimination of the customary reheating process between cogging and finishing was made possible.

A type of mill which enables work to be achieved on all sides of a work piece without the use of grooved rolls was first used in Germany about 1848. Known as the Universal mill it incorporated vertical as well as horizontal rolls and in addition to rolling flanged beams, girders and other sections is also able to roll plate and sheet. Improvements to the Universal mill culminated in Grey's reversing mill for broad-flanged beams.

The earliest record of a continuous mill was that devised by G. Bedson, manager of a wire-drawing works in Manchester. Bedson's continuous mill for wire rods patented in 1862, had originated from seeing cotton fibres at various Manchester textile works being drawn through successive rollers. In Bedson's mill there were sixteen stands of rolls alternately horizontal and vertical; but instead of the diameter of the rolls being adjusted to the increasing speed of the bar going through, the speed of each succeeding pair of rolls was accelerated. Their successive synchronization was a technical triumph, for if an error occurred, the rod emerging from the mill at a final speed of about 1,000 ft. a minute would loop itself around the rafters of the building, causing the workmen to 'dive for the exits'.* A. S. Hewitt, the United States commissioner to the 1867 Paris Exhibition, described the Bedson mill as 'the most remarkable specimen of rolling in the Engineering Department of the Exhibition'.

In 1892 a continuous mill for rolling strip was erected at Teplitz in Austria. On this mill 8 in. thick steel slabs weighing 1,000 lb. were hot-rolled to 0·28 in. thick on three high mills located side by side, finishing on five-stand two-high mills in tandem, the strip emerging from the final stand at a speed of 400 ft. per minute and a gauge of 0·11 in. Because of various technical difficulties connected with lack of uniformity in the thickness of the material no real success attended the effort of these pioneers and the mill was closed some years later.

In 1902 the American Tinplate Company installed in their Monongaheta Works a mill designed by the engineer, C. W. Bray. It consisted of eight two-high mills with rolls 26 in. diameter by 32 in. width, stands in tandem working on bar material for the production of tinplate. Continual roll breakage and high scrap loss prevented the mill from operating economically and it was abandoned in 1905. No real progress was achieved until 1926 when the Columbia Steel Co. erected a plant at Butler, Pennsylvania, to roll strip up to 48 in. wide. It consisted of a two-high reversing universal roughing unit with rolls 27 in. diameter and 48 in. furnace width, followed by four finishing four-high mills with working and support roll diameters of 16 in. and 22 in. respectively. Slabs 2½–5 in. thickness were given 5–7 passes in the Universal mill which reduced the thickness to between ½ in. and ¼ in., the material then passing to the finishing stands, emerging from the last stand 0·05 in. thick. This mill became the prototype of the present-day continuous wide-strip mill.

From experience gained at this and other plants it was found that successful continuous rolling of wide strip depended not only on the shape, contour, temperature and deflection of the rolls, but also on the shape, composition and temperature of the material being rolled.

By the beginning of the 20th century the steam engine was being displaced by the electric motor. Although more efficient, reliable and flexible, enabling separate drives to each of the

---

* *History of the British Steel Industry*, Basil Blackwell & Mott Ltd., Oxford.

rolls, the motor did not have matters all its own way. As late as 1912 the economic balance was still delicate, and although there were about 30 electric rolling mills operating, and some 20 more building in Britain and on the Continent, steam engines still had many supporters on the grounds of both cost and simplicity. In Germany steam driving was still cheaper in 1907, and in 1911 it was being reported that some of the German works which had electrified were reverting to steam driving on grounds of economy. In the United States, electric driving did not arouse much interest until about 1906, but development after was rapid and by the end of 1911 there were about 50 electrically driven mills in operation.

EXTRUSION

Extrusion is essentially an operation of producing useful metal shapes such as rods, tubes and various solid and hollow sections by forcing hot solid metal through a suitably shaped die by means of a hydraulic ram. The operation can be likened to the squeezing of tooth paste from a tube and is commonly carried out hot when the metal is soft and plastic, but is also in some instances carried out in the cold. When tubes are required, the centre of the die orifice is partially closed by a mandrel so that the metal is forced to flow through the annular space between the mandrel and the die. The method is commonly used in the manufacture of rods, tubes and sections in such metals as lead, copper, aluminium, magnesium and their alloys and has recently been applied to steel.

It is perhaps natural that the earliest work of this nature should have been with the soft metal lead, and that up to the end of the 19th century, this was practically the only metal so worked. The first operator on the scene appears to be Joseph Bramah (1748–1814), inventor of the Bramah hydraulic press, who in a patent granted in 1797 described a machine 'For making pipes of lead of all dimensions and of any given length without joints'. To achieve this object lead was maintained in a molten condition in a pot and forced through a horizontal tube projecting from the pot by means of a hydraulic press. A mandrel was supported axially in the tube, the lead passing through the intervening space being cooled as it emerged in the form of pipe. Lead tubing at this period had been produced by casting a hollow cylindrical billet and then lengthening this into a long tube by inserting a mandrel and rolling. Alternatively sheet lead was bent over a mandrel and joined with a soldered seam. Incidentally, this process is still used in the case of pipes over 12 in. in diameter.

The first practical machine working on the extrusion principle was constructed in 1820 by T. Burr, a Shrewsbury plumber. The press differed from Bramah's in that the operation was carried out on the metal enclosed in a container integral with the press, a hydraulic ram forcing the lead out through the container die to form the pipe. The molten metal was poured into the container and allowed to solidify before extrusion, the extrusion temperature being in the region 200–250° C.

By the middle of the 19th century the size and power of the extruding press was gradually increasing, and the process was being extended to the sheathing of underground electric cables to protect them against corrosion. This was first accomplished by G. Borel in France in 1879, lead tubing being extruded and shrunk over the insulated copper cable. The machine had the disadvantage of not being continuous, a fresh billet of lead having to be inserted every hour or so. Huber in Germany a few years later remedied this defect by devising a machine having two hydraulic rams acting in horizontal cylinders opposite each other, the cylinders being continuously supplied with lead from a melting pot set in a furnace on top of the press. The rams moved forward to force the lead through the die block located between them as a sheath over the cable. By this means long lengths of cable could be continuously sheathed. The machine found a wide application and large units up to 5,000 tons pressure were in use until comparatively recently they were superseded by a vertical type of cable press which had its origin in America.

In the modern cable-sheathing machine, molten lead is poured into a vertical cylinder and allowed to set. A ram then descends in the cylinder and forces the hot but solid metal through a port in the bottom into the cable box. This is a container communicating with the cylinder of the lead press and provided with a port at its side through which the lead emerges. The cable passes through the cable box horizontally and out through the same exit port, the lead being forced to occupy the annular space between the cable and the sides of the die. Presses range from 300–4,000 tons pressure capacity, exerting an extrusion pressure of 25–28 tons per sq. in. The sheathing of cables with metals other than lead has in recent years received some attention. Extrusion of aluminium on to a cable core has been the subject of research for many years, and was first developed along lines similar to the sheathing of cables with lead. An early difficulty was the high temperature necessary for extrusion. Some of the first attempts were made at 450–500° C. The first satisfactory presses were those operated by Siemens-Schuckertwerke after World War Two. Improved designs were developed and put into service in 1952 using billets which were heated by induction to 300–400° C. These machines were produced by Hydraulik G.m.b.H., Duisburg, and operated vertically. The most recently developed Schloemann press operates horizontally and is designed expressly for extruding aluminium. It uses two opposing rams operating simultaneously and extruding 99·5% pure billets. It is capable of sheathing cable of any diameter from under a quarter inch to four inches, and the cooling system ensures that no damage to the cable is caused during the normal operation of stopping the machine for recharging with billets.

With the achievement of successful lead extrusion many efforts were made to adapt the process to other metals such as copper and brass. The difficulty, as compared to lead extrusion, was that a much higher temperature and pressure was required to render the metals sufficiently plastic to undergo successful deformation. Dies and containers had to be provided to withstand the high thermal and stress conditions.

G. A. Dick (1838–1903) deserves the major credit for overcoming the many obstacles involved. Dick's first patent for an extrusion press was obtained in 1894 and was followed by several others; present-day presses and processes of extrusion embody many of the devices originated by Dick. He was born in Germany, but settled in England in 1870 and after numerous preliminary experiments with a small hydraulic press was eventually successful in 1894 in extruding brass and other copper alloys in round and solid sections and in tube form.

The modern extrusion machine is generally a horizontal hydraulic press of massive proportions and power capable of exerting up to 20,000 tons ram pressure and handling metal billets of 36 in. diameter and 76 in. long. It comprises a heavy steel container fitted with alloy steel liners to take the wear and thus conserve the strength and tightness of the structure. The die which fits into a recess in the face of the die holder is composed of some hard material such as sintered carbide or chrome nickel steel. For the production of tubes a mandrel is attached to the end of the ram, and passes through the hollow metal cast billet entering the die before the ram puts pressure on the metal, which is then forced out through the annular space in the form of tube. The whole massive structure is supported on columns embedded in extensive foundations.

Although the extrusion of aluminium and magnesium alloys, the cupro-nickels and nickel silver necessitates that the container be heated to 300–350° C., the brasses do not usually require any additional heating, the hot brass billet retaining sufficient heat to be successfully extruded.

The application of the extrusion process to steel was held back for many years because of difficulties concerned with the seizure of the metal in the die, a tendency which is common with high melting-point metals. The first commercial production was achieved at the Mannesmann works in Germany, small-bore tubes being produced about 1937. A billet heated to 1200–1300° C. was placed in the chamber of a mechanical press and extruded in 40–50 ft. lengths in diameters of $\frac{3}{8}$–1 in. It was found that a high rate of extrusion was necessary in order to arrest the excessive wear and tear of dies and mandrels, and also to obviate seizure due to the high friction. In this connection the introduction of glass as lubricant in the late 1940's by the French inventor, Jacques Segournet at the Persan plant of Industrial D'Etirage et Profilage de Métaux was of the highest importance. The lubricant is applied by rolling the billet in powdered glass or glass fibre or wool before it is put in the chamber; the glass melts in contact with the hot billet and flows to the die, thus lubricating its surface and reducing friction to a minimum.

WIRE DRAWING

For centuries wire was formed by hammering metal to a flat plate from which narrow strips were cut and then rounded by beating and rotating between two flat surfaces. Later thinner wire was formed by being drawn, or pulled through hard iron plates known as 'whortle plates' containing holes rather smaller than the original diameter of the wire. The holes in the whortle plates wore out of size very quickly and had to be reset after each operation. This was done by hammering the back of the plate until the hole partly closed up and then repunching it with a suitably sized tool. When one cavity in the plate was worn past repair a new hole was punched. The method was dependent upon the strength of the drawer in pulling the wire through the cavities in the whortle plates and hence was limited to the softer metals and it was not until the advent of the steam-powered engine in 1769 that manual drawing was eliminated and the drawing of other metals was rendered possible.

With the development of the telegraph in 1844, the invention of barbed wire in 1867 and the telephone in 1876, the demand for large tonnages and long lengths of steel (and copper) wire became acute and led to the mechanization of the drawing process. Supply of the raw material was made possible by the invention of the Bessemer steelmaking process in 1856 and the open-hearth method a few years later, coupled with the development of continuous mills for the manufacture of wire rods by G. Bedson in 1862 at Richard Johnson & Co.'s Manchester wire-drawing works.

One of the highlights in the mechanization of the wire drawing operation was the development of a power-driven revolving drum or block, arranged to pull the wire through the die clamped to a bench. The wire rod in coil form was placed on a freely rotating frame or swift, the wire-drawer pointing one end of the rod and pushing it through the die. The point protruding from the die was then secured to the block which was then revolved, the rod being drawn through the die, the wire being taken up on the vertically driven drum or block. By this type of operation only one pass was possible and when the wire had to be drawn more than once, the coil had to be taken off the block, put back on the swift, and the die changed for one of the next size. This led to the development of the multi-block machine patented by S. H. Byrne in 1885, in which several machines were placed one in front of the other, each block acting as a swift for the next one. In operation the wire was drawn through a first die, coiled around a block, then passed through a second die, and the same procedure repeated a number of times using successively smaller-holed dies. As the wire gets thinner after each pass through a die it becomes proportionately longer. Consequently each block had to revolve faster than the previous one. The problem of matching the speed of the blocks caused quite a problem in machine design. One solution was to provide a bobbin at the top of the block mounted on a floating ring; the wire passing round it and over an overhead pulley to the next die. If the succeeding block is drawing faster than the preceding one, the bobbin at the top of it will move in a certain direction and wire will be unwound from it. If the succeeding block is drawing less wire than the preceding one, the bobbin will move in the other direction and wire will accumulate on this drum.

Continuity was achieved by butt welding the end of one coil of rod to the beginning of the next one.

The original iron or steel dies were later made of a hard steel containing 2% carbon and 4% chromium, but wire drawing dies are now usually made of the much more wear-resistant tungsten carbide, and for very fine wires diamond dies are in use. In order to prevent seizure between the wire and the die, a lubricant in the form of powdered soap is placed in a compartment behind the wire-drawing die, the rod or wire passing through this box before entering the die. Cooling water is also caused to pass through the die box in order to dissipate the frictional heat which is generated.

POWDER METALLURGY

The traditional methods of shaping metals by casting or by hot or cold working are difficult and sometimes impossible to apply to many metals. Such refractory metals include tungsten (m.-p. 3380° C.), molybdenum (2622° C.) and tantalum (2996° C.) whose melting-points are too high to enable them to be melted by conventional means. In these cases an alternative procedure has emerged based on the fact that metals in powdered form may be caused to adhere together without being melted by employment of high pressure. The technique known as powder metallurgy consists in subjecting the powdered metal contained in a mould or die of the shape desired to a high pressure followed by sintering at a suitable temperature. The method provides either finished metal components or compact blocks of metal for subsequent mechanical working. Having been applied first to the refractory metals the method has been extended to many of the more tractable metals.

The technique originated more than a century ago. The German chemist, G. Osann, when determining the atomic weight of copper by the reduction of its oxide, noticed that the reduced metal sintered to a compact mass. In a subsequent paper he gave full details of the process and suggested that the method could be used for making impressions of medals and the like from powdered copper, silver and lead metals.

The preparation of compact platinum from the then infusible metal by W. H. Wollaston in 1830 represents one of the earliest applications of powder metallurgy. Before Wollaston's work, owing to the failure to provide the high temperature (1770° C.) involved in the melting of platinum, little use could be made of the metal. Starting from a finely divided platinate precipitate, Wollaston heated to a loose spongy product, pressed the sieved powder to briquette form and then reheated to 800–1000° C., obtaining a compact metal mass which could be hot forged into a malleable ingot for subsequent rolling into sheet metal. Although the subsequent discovery by Ste-Claire Deville in 1859 of the oxy-coal gas flame capable of melting platinum eventually replaced Wollaston's procedure, the purest platinum is still prepared by powder metallurgy in order to avoid contamination during melting.

Tungsten (m.-p. 3380° C.) is another metal which was developed through the advent of powder metallurgy. As obtained by reduction of its oxide, tungsten is a grey, brittle powder. Because of its high melting-point it had for long been recognized that the metal would be eminently suitable for electric lamp filaments, but because of its brittleness attempts to draw the metal resulted in failure.

The first metal filaments used in electric lamps consisted of osmium introduced by Dr. Welsbach of Austria in 1898. The filament was produced by extruding a mixture of the metal powder with an organic binder such as gum and then volatilizing the binder by heating in an atmosphere of hydrogen to prevent oxidation. Sintering of the fragile wire so obtained *in vacuo* inside the lamp bulbs completed the procedure. The process was applied to tungsten but lack of ductility was a serious shortcoming. This was to some extent overcome at the Siemens-Halske Werke, Berlin, in 1908. Tungsten powder was mixed with nickel oxide and pressed into rods. The compacts were then sintered in hydrogen at 1000° C. followed by volatilization of the nickel at 1580° C. By alternate heating and rolling the rod can be worked to a small diameter filament. The next development was in America in 1909, when W. D. Coolidge at the General Electric Co. discovered that by alternating a mechanical hammering process, termed 'swagging', with heating, the ductility of the metal could be increased until finally it could be drawn into wire at room temperature.

Production of powder of the requisite properties is an important stage in the procedure of powder metallurgy. Powders of metals and alloys may be produced by mechanical methods such as grinding, machining and milling; other metal powders can be obtained by reduction of the metal oxide by hydrogen or carbon. Copper, iron, cobalt, molybdenum and tungsten can all be so prepared in powder form from their oxides. Electrolysis is also used, and aluminium, tin and lead are transformed into powder form by atomization, molten metal being poured through an orifice into a chamber and sprayed with a high pressure jet of inert gas, the instantaneous chilling converting the metal into a finely divided dust.

At the present day powder metallurgy is mainly used in making large numbers of identical components usually of relatively small size, such as permanent magnets, coins, medals, small gear wheels and brushes for motors and dynamos. A novel extension of its application is in the manufacture of the oil-less bearing which can be impregnated with oil and made self-lubricating. Such bearings are designed to retain within their structure a sufficient amount of oil to last for several years. Their invention in 1870 stems originally from the work of S. Gwynn of New York, whose patents covered the pressing with lubricants of metal powders such as tin, zinc, lead and bronze. In 1910 E. G. Gilson described the production of bronze material containing graphite and oil by heating copper oxide, tin oxide and graphite, the oxides being reduced to bronze metal which could hold as much as 40% volume of graphite. Since the heating is carried out at temperatures below the melting-point of bronze, the sintered compact is porous and spongy in character. In service the rotating shaft of the mechanism on which the bearing is installed causes a rise in temperature which initiates oiling from the bearing; on cooling capillary action of the porous metal causes reabsorption of oil. The degree of porosity may be varied and controlled from 10 to 40% by volume. The bearings are impregnated with oil by a vacuum treatment and this initial oiling is usually sufficient to last the lifetime of the machine in which it is installed. Self-lubricating bearings based on other metals such as iron and iron-copper are also in production. The largest user is the automobile industry; others include refrigerators, electric clocks, electric motors, vacuum cleaners, and washing machines.

BIBLIOGRAPHY

1. Institute of Metallurgists. *Metals in the Service of Mankind*, 1950.
2. LARKE, E. C., *Rolling of Strip Sheet and Plate*, 1957. Chapman & Hall Ltd., London.
3. PEARSON, C. E., and PARKER, R. N., *Extrusion of Metals*, 1960. Chapman & Hall Ltd., London.
4. CARR, J. C., and TAPLIN, W., *History of the British Steel Industry*, 1962. Basil Blackwell, Oxford.
5. British Iron & Steel Federation. *Guide to the Finishing Processes of the Iron and Steel Industry*.
6. JONES, W. D., *Fundamental Principles of Powder Metallurgy*, 1960. Edward Arnold, London.

CHAPTER NINE

METALLOGRAPHY

ETALLOGRAPHY deals with the study of the structure and constitution of metals and alloys and their relation to the physical and mechanical properties. Although formerly confined to the microscopical examination of metals it now comprises all methods employed in the study of the internal structure of metals. Comparatively little was known about the metallic state until the latter half of the last century, though an important landmark on the way was the observation in 1808 by Alois von Widmannstätten of Vienna of the remarkable crystalline structure revealed by a polished and etched surface of a meteorite—the Agram (Zagreb) iron which had been observed to fall in 1751. The characteristic mesh-like arrangement has come to be known as the Widmannstätten structure.

The science of modern metallography may be said to have been introduced by Dr. H. C. Sorby (1826–1908) of Sheffield, who in 1849 founded petrology by using the microscope for examining rocks in thin sections by transmitted light. He was led from this study to that of iron and steel. Metal grains are, however, opaque even in thin sections, their optical properties not being readily measured as in the case of rock crystals, which can transmit light. Instead the surface of the metal has to be illuminated in such a manner that light is reflected from it into the objective of the microscope. Sorby solved the problem by the provision of reflectors positioned so as to throw the light directly down on the object and from this reflected through one half of a microscopic lens in such a manner that a polished surface appeared bright, and a rough surface comparatively dark.

An important part of Sorby's contribution was his polishing technique secured by the use of rubbing with a sequence of emery papers commencing with a coarse paper and finishing with the smoothest. All trace of roughness was then removed by means of rouge and water. Etching was usually accomplished by dilute nitric acid which served to differentiate the various constituents by differences in the light reflecting power of the metal surface. Finally, thin glass covers were mounted over the surface with Canada balsam.

By his technique Sorby made possible the identification for the first time of the following basic metallographic constituents present in steel:

(*a*) Free iron.
(*b*) Iron combined with carbon.
(*c*) A pearly lamellate constituent (formed from *a* and *b*).
(*d*) A hard brittle uncrystallized constituent.

Following a suggestion in 1888 by the American metallurgist, Dr. H. M. Howe, these constituents were renamed ferrite, cementite and pearlite. He also suggested the term sorbite for a nodular constituent formed by the tempering of steel at 400° C. The term has now, however, practically disappeared from present-day nomenclature.

Ferrite was derived from the Latin *ferrum*, meaning iron. Cementite, the carbide (Fe_3C) of iron, was given its name by Howe, because it was first observed in cemented or 'blister' steel. The pearly constituent of Sorby was so named because a sample of steel containing the material took on, after polishing and etching, a mother-of-pearl lustre. Howe, for the sake of brevity, suggested that it be called pearlite.

Sorby's hard brittle uncrystallized constituent was named 'hardenite' by J. O. Arnold, professor of metallurgy at Sheffield, because of its intense hardness. This term has also fallen into disuse being replaced by 'martensite'.

Sorby not only identified many of the microscopic constituents of iron and steel, but was able to relate the properties to specific visible structural changes. He showed that metals were undoubtedly crystalline, and recognized that as carbon was added in increasing amounts to iron, a sequence of constituents appeared and that the pearly constituent of steel was itself composed of lamellae of iron alternating with a compound of iron and carbon.

The effects of quenching on the structure of steel also attracted his attention and he correctly attributed the hardness of steel after quenching to suppression of the decomposition of

322

pearlite, the steel retaining properties intermediate between those of soft iron and the intensely hard brittle compound (cementite).

On raising the temperature so as to temper the steel the two constituents (ferrite and cementite) separate so as to give rise to a structure similar to that when the steel is slowly cooled. These ideas on the hardening of steel were given in a paper 'On the Microscopical Structure of Iron and Steel' read to the Iron and Steel Institute in 1885, and to the Yorkshire Geological and Polytechnic Society in 1886.

Many of Sorby's photomicrographs of iron and steel are preserved at the University of Sheffield and testify to the excellence of his technique.

A. Martens, director of the Königlich Technische Versuchsanstalt, wrote a large number of papers describing the structure of various ferrous materials such as cast iron, malleable iron and spiegeleisen (ferro-manganese). In 1880, in collaboration with the Carl Zeiss Company, he designed equipment for metallographic photo-micrography which formed the basis for present-day apparatus.

F. Osmond (1849–1912), when employed at the Creusot steelmaking plant in France, was concerned with the effect of heat treatment on the structure of steel. In collaboration with J. Werth he made extensive studies of the role of carbon in heat treatment, and by combining microscopy with thermal studies made many advances. In 1887 he showed that if iron be cooled from its melting-point, three breaks or arrests occur in the cooling rate due to the evolution of a small amount of heat whereby the rate of cooling is retarded. He likewise found that on heating a small amount of heat is absorbed by the metal, the arrests occurring at a slightly higher temperature than that on cooling. These critical temperatures or points were represented by Osmond by the letter A* from the initial letter of the French word *arrestation*.

The arrests on cooling were symbolized Ar, where r is the initial letter of the word *refroidissement*, meaning cooling, those on heating being represented by Ac where c is the initial letter of the French word *chauffage*, meaning heating.

---

* Professor D. K. Tschernoff, the Russian metallurgist, had previously in 1868 used the letter 'A' for the temperature at which hardening by rapid cooling becomes suddenly possible in high carbon steel.

The approximate temperatures at which these Arrests were found to occur were as follows:

Heating: Ac_4 1404° C. Ac_3 910° C. Ac_2 770° C.

Cooling: Ar_4 1400° C. Ar_3 890° C. Ar_2 760° C.

Osmond suggested that these arrests represent changes from one allotropic form of iron to another. This indicates that there are four allotropic states of iron. The form stable at ordinary temperature and up to 770° C., i.e. A_2 is called alpha iron; the form stable between 770° C. and 890° C. is called beta iron; that between 890° C. and 1404° C. gamma iron; and that between 1404° C. and the melting-point of iron, delta iron.

Influence of Carbon. Osmond further showed that if iron contains a small proportion of carbon, a new critical point appears at approximately 690° C. or 710° C. This is known as the Ar_1 arrest, it being inferred that the carbon is held in solid solution by the iron at a temperature above the Ar_1 arrest, but at that temperature the carbon separates from solid solution in the form of cementite. As the proportion of carbon increases, the Ar_3 arrest is lowered and joins up with the Ar_2, the double $Ar_{2\ 3}$ point then descends and finally coincides with the Ar_1 point when the proportion of carbon has attained 0·8–0·9%. The significance of these observations was that carbon retards the transformation of gamma iron into alpha iron, thereby increasing the stability of gamma iron. Cooling or quenching of the iron carbon alloy may result in gamma iron persisting at ordinary temperature. Alloying with say manganese, nickel, chromium or tungsten considerably lowers the temperature of the Ar_1 arrest point so that gamma iron may persist when the iron is cooled down in air to room temperature. This solution of carbon in gamma iron first noted in 1895 was named Austenite by Osmond after Roberts-Austen, the name first appearing in print in 1901.

Osmond also made investigations into the structure of the hard acicular constituent named by Arnold hardenite, obtained by the rapid cooling of steel. He renamed this structure martensite after A. Martens. After showing the sequence of structural changes on steels quenched from progressively higher temperatures he produced a diagram illustrating how cementite decomposes above the Ac_1 point to yield carbon which diffuses into ferrite to form martensite. The area occupied by martensite increases with temperature or by increasing the carbon content. Rapid cooling preserves the carbon as hardening carbon and Osmond believed this carbon served to retain some of the iron in the beta form. He found that martensite is best formed when a steel containing $0 \cdot 2 - 0 \cdot 8\%$ carbon is cooled from above the temperature of the A_3 arrest slowly to the A_2 arrest and then suddenly quenched in a freezing mixture at $-20°$ C.

Osmond's theory that hardening was due to beta iron was resisted by Howe, who maintained that the hardness of steel depended on the amount of carbon present in it and the form in which carbon existed, which varied with heat treatment. The argument inspired a great deal of experimental work which eventually led to the modern conception that gamma iron changes to alpha iron without any deposition of carbon, the structure being considered as a supersaturated solution of carbon in alpha iron.

Osmond was a brilliant metallurgist who, by combining microscopy and thermal analysis as a means for studying the structure of steel, pioneered investigations into the complicated relationship between iron and carbon, which largely determines the properties of steel and how it behaves when heated and cooled.

Sir W. C. Roberts-Austen (1843–1902), after studying at the Royal School of Mines, was appointed a chemist at the Royal Mint in 1865 and was led to the study of steel and microscopy through investigations concerned with alloys of gold and silver. For these thermal investigations he had to use the slow and cumbersome calorimetric methods of his day. H. Le Chatelier, at the École des Mines in Paris in 1887, developed the thermo-electric couple, the temperature being indicated by the movement of a ray of light reflected from the mirror of a galvanometer on to a graduated scale. This pyrometrical method was far in advance of the inaccurate calorimetric methods then in vogue and Roberts-Austen immediately took it up and was able to obtain evidence as to the composition of the eutectics in alloys of high melting-point.

At the end of the century there was still some confusion as to the effects of time and temperature upon the phases present in an alloy. Based on a long succession of investigations which had been conducted on a series of iron-carbon alloys with a progressive variation of carbon, Roberts-Austen in 1896 produced the first iron carbon constitutional diagram (Fig. 23). The diagram was constructed by plotting temperature as ordinate and percentage of carbon in the alloy as abscissa, and showed the range in composition and temperature within which the phase changes are stable, and also the boundaries at which the phase changes occur. Such a diagram, in addition to establishing a correlation between the microstructure and properties of

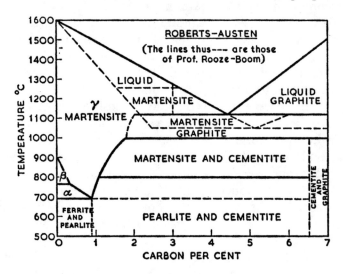

FIG. 23

FREEZING-POINT CURVES OF THE CARBON-IRON SERIES
Source. Roberts-Austen, *A Record of his Work* (p. 44)
S. W. Smith (1914)

plain carbon steels was basic to an understanding of heat treatment principles, and so proved a new basis for metallurgical concepts. Osmond used the new diagram to explain the successive changes in steel structure he had previously reported, and H. W. Roozeboom, the German scientist, used the diagram to draw his conclusions as to the application of the Phase Law and general thermo-dynamic principles.

He maintained a close friendship with Osmond and in conjunction with him communicated a paper to the Royal Society in 1896 on the microstructure of gold containing certain impurities which affected its suitability for coinage. Based on a similarity which he found to exist between the action of impurities in gold and iron, he initiated experiments on the role of carbon in iron and from these investigations resulted the iron-carbon constitutional diagram. He eventually succeeded Dr. Percy as professor of metallurgy at the Royal School of Mines.

X-ray Crystallography. Microscopical examination is limited to the surface of metals, and what is seen in only a two dimensional section of the solid structure. Visible light has wavelengths of the order of 10^{-4} cm. and for examination of details of the internal structure much shorter wavelengths are essential. The discovery by W. C. Röntgen in 1895 of X-rays having wavelengths of the order 10^{-8} cm. opened up the way to the determination of the arrangement of the atoms in a metallic crystal, for the wavelengths of X-rays happen to approximate the same distance between the atoms in a crystal.

Solid crystalline matter serves as a three-dimensional grating of the requisite magnitude for X-ray diffraction by virtue of the regular arrangement in space of the atoms just as closely ruled lines on a glass or metal plate serve as a diffraction grating producing a spectrum of ordinary light. The discovery in 1912 that under certain conditions X-rays are diffracted by crystals and that the pattern of the diffracted X-ray intensities could be obtained on a photographic plate was due to Professor Von Laue of Zürich. The phenomenon was given expression in the now familiar Bragg's Law:

$$n\lambda = 2 \, d \sin \theta$$

where n is an integer, 1, 2, 3, etc. (the order of the spectrum), λ is the X-ray wavelength, d is the interplanar spacing or distance between two identical planes of atoms in the crystal, and θ the angle of incidence of the X-ray beam on this set of planes.

The application of the X-ray diffraction method to metallography resulted largely from the further work of Debeye and Sherrer (1916) in Europe and A. W. Hull (1917) in America who found that when metal crystal powders were exposed to monochromatic X-rays (i.e. beams consisting of one wavelength only) a sufficient number of crystals would be in exactly the requisite position to reflect the rays and thus afford an image of the atomic arrangement and hence the structure of metals.

The pattern could be used to identify the solid and also the actual condition in a specimen in terms of grain size, effects of heat treatment, fabrication, fatigue, etc.

Although X-ray diffraction is useful for regular crystal lattices, it has limitations in that a direct image of the atomic pattern cannot be focused directly as in the optical instrument, the image of the structure having to be computed from the diffraction pattern observed, which is a complex and lengthy proceeding.

Radiography. Of metallurgical interest is the use made of X-rays to penetrate metal opaque to ordinary light. Intensity of the rays during their passage through a metal decreases, each inch reducing approximately the intensity by half. Thus if the object is a casting with a wall thickness of say 2 in. containing a blowhole 1 in. in diameter, the rays passing through the blowhole will be twice as strong as the beam passing through the same portion of the casting. When the rays strike a photographic plate or film they will affect it in proportion to the intensity, any non-uniformity thus being shown up. The rays that have come through the blowhole, having been less absorbed, will make a spot on the film that is blacker than the area representing sound metal. The black spot will in fact be the shadow of the blowhole. Castings, forgings, welds, etc., may thus be inspected for blowholes, cracks, shrinkage, cavities and slag inclusions.

Electron Microscopy. The resolving power of the optical microscope is limited by the effective aperture of its lens system and the wavelength of light used for making the inspection. Thus no matter how much the size of an optical image is increased there will always be a limit beyond which magnification fails to reveal any further details of structure. This limit for optical magnification is of the order of $1-2$ μ whereas microstructure can be said to cover the range $0 \cdot 01$ to 100 μ (10^{-6} to 10^{-7} cm). Although X-ray microradiography can be very useful in the lower part of the range, i.e. below 1 μ, there is at present no microscope in the usual sense available by which the resolving power of X-rays can be utilized.

Two discoveries in the 1920s materially contributed to the development of an instrument capable of resolution of structure in the lower part of the range, namely the electron microscope. The first, that very short wavelengths are associated with an electron beam, was due to Louis de Broglie in Paris, Dawson and Gerner of the Bell Telephone Co., U.S.A., and G. P. Thomson at Aberdeen.

The second was the demonstration by H. Busch in Germany in 1926 that suitably shaped magnetic fields could be used to produce an image of an electron beam and that an enlarged image of the source could be obtained. Following up these discoveries in 1931, M. Knoll and E. Ruska in Germany built the first electron microscope which followed in general outline the technique applied to the construction of a high-voltage cathode-ray tube, the electrons being produced from a cold aluminium tube situated at the top of the instrument, the magnetic objective lens providing an image on a fluorescent screen. The images produced by this early microscope were crude and inferior to those obtained by an optical instrument; but continued research led to improvements and by 1940 the electron microscope was being manufactured commercially by Siemens in Berlin. In the modern instrument the electron beams are produced by a hot-wire cathode and accelerated through an anode by a potential of 50,000 to 100,000 volts. The electrons pass through the specimen by which they are selectively absorbed or diffracted. The beam is then magnified by the objective magnetic coils and projected on a flourescent screen or photographed. The differential transmission of electrons through the specimen is the source of the detail of the image. The advantages over the optical microscope are (1) increased resolving power enabling much finer resolution of detail, (2) greatly increased depth of focus enabling magnification in three dimensions.

American Metallographers. Dr. H. M. Howe has already been mentioned in connection with the nomenclature of the microstructure in steel. He did little research himself, but he had the ability to seize upon the significance of other people's work and to synthesize both science and practice into real understanding which he clearly expressed.* He was critical of Osmond's work,

* C. S. Smith, *A History of Metallography.*

the basis for much of his criticism being provided by investigations by Dr. A. Sauveur, then employed by the Illinois Steel Co. Having heard of the interesting results obtained by Sorby and Osmond, Sauveur began in 1891 to use the microscope in the study of steel rail manufacture, publishing his results at a meeting of the American Institute of Mining Engineers in 1893; and a further paper on the mechanism of hardening of steel appeared in 1896. In the same year he founded a journal, *The Metallographist*, to further the new science. He may be said to be one of the first to carry out metallography in a quantitative sense.

In more recent times attention may be drawn to the work of E. C. Bain and E. S. Davenport of the United States Steel Corporation in connection with the construction of isothermal-transformation diagrams. When a steel is quenched from a temperature at which it is austenitic to a temperature at which austenite is no longer the stable phase, it transforms ultimately to a mixture of ferrite and carbide or to martensite. The time taken for transformation to ferrite and carbide at each of a series of temperatures can be represented by a diagram relating temperature, time and progress of transformation. Diagrams of this type are known as isothermal-transformation diagrams and they have for many years made a contribution to the understanding and control of the heat treatment of steels, providing a graphic forecast of the behaviour of the steel during heat treatment. The most direct of the commonly used methods of obtaining data for these diagrams is the microscopic method originally used by Bain and Davenport in 1930. This involves austenitizing small samples of steel at an appropriate temperature, quenching to a sub-critical temperature, holding at that lower temperature for progressively increasing times and finally quenching to room temperature. By microscopic examination of the quenched samples, it is thus possible to determine the time taken for the transformation to start, the rate at which it proceeds, and the time required for its completion. Though reliable, the method is time-consuming, and skill is required to interpret the large number of microstructures which must be examined to provide an accurate diagram. It is consequently more often convenient to follow the course of the transformation by recording the changes in some physical property accompanying transformation, such as dilation, magnetic permeability, electrical resistance and changes in the intensity of the line in X-ray diffraction patterns. Since breakdown of austenite is accompanied by expansion, a change of length is readily measured and hence the dilation method is by far the most widely adopted.

BIBLIOGRAPHY

1. SMITH, C. S., *A History of Metallography*, 1960. University of Chicago Press, Illinois, U.S.A.
2. SMITH, S. W., *Roberts-Austen, an account of his work*, 1914. C. Griffin & Co., London.
3. OSMOND, F., *Méthode générale pour l'analyse micrographique des aciers du carbone*. Bull. Société d'Encouragement pour l'Industrie Nationale, Paris, 1895. 10, 480–518.

GLOSSARY

ACID STEELMAKING PROCESS   A steelmaking process operated in a
Bessemer, open-hearth or electric furnace in which the furnace
is lined with an acid refractory, e.g. silica, and for which pig
iron low in phosphorus is required as this element is not
removed.

ANNEALING   General term denoted by heating followed by cooling
at a suitable rate with the object of (a) softening a metal
hardened by cold working, (b) removing internal stress, (c)
refining the grain size.

AUSTENITE   Is the solid solution of carbon in gamma iron formed
when steel is heated above 723° C. Normally does not exist
below this temperature but may be preserved at normal
temperature by addition of certain alloying elements.

BASIC STEELMAKING PROCESS   Steel produced in Bessemer, open
hearth or electric furnaces lined with a basic refractory such as
dolomite or magnesite, a slag rich in lime being formed which
removes phosphorus.

BLAST   Air under pressure blown into a furnace through tuyères.

BLOOM   (a) An intermediate product formed in the rolling of steel,
usually square in section and more than 5 in. square, smaller
sizes being known as billets. (b) Iron that has been produced
in a solid condition as a result of smelting iron ore.

BOSH   Tapering portion of a blast furnace between the bottom of
the stack and the top of the hearth.

BRITISH THERMAL UNIT (B.Th.U.)   Amount of heat required to
raise the temperature of one pound of water through one degree
Fahrenheit.

BUDDLE   A circular convex table formerly used for the concentration of tin and other mineral slimes in Cornwall and elsewhere.
A revolving distributor supplies the finely divided ore in water
to the upper surface of the table, heavy mineral being built
up (buddled) on the upper portion, the lighter gangue material
flowing down over the perimeter into a launder.

BURDEN   The material charged into a blast furnace, more specifically the ratio of the total weight of ore and flux to the weight of
fuel.

CALCINE   The product resulting from the calcination or roasting
of ore.

CAST IRON   An iron alloy containing substantial amounts of carbon in the form of cementite or graphite which renders it brittle and non-malleable and hence unsuitable for forging and rolling.

CEMENTATION   (*a*) Process by which wrought iron is converted into steel by heating in contact with carbonaceous material. (*b*) Raising the carbon content of steel by impregnation with carbon, also known as case hardening. (*c*) Precipitation of metal by means of another metal from solution resulting from leaching, e.g. copper precipitation by means of iron.

CEMENTITE   A hard brittle compound of iron and carbon (Fe_3C) occurring as a constituent of cast iron and steel.

COGGING   The operation of rolling an ingot to reduce it to a bloom or billet.

CORROSION   Slow destruction of metals by chemical or electro-chemical attack.

CONSTITUTIONAL DIAGRAM (Phase or Equilibrium diagram)   Graphical representation of the equilibrium temperatures and compositions within which the different phases or constituents occurring in an alloy system are stable.

CUPOLA   A small shaft furnace used for the melting of pig iron for castings in foundries.

DECARBURIZATION   Removal of carbon from an iron alloy, usually steel, by heating in a medium which reacts with the carbon.

DEOXIDATION   The process of elimination of oxygen from molten metal before casting by addition of substances which abstract oxygen by formation of oxides.

DIFFRACTION   Appearance of alternating dark and light-coloured bands or fringes when light is deflected by passing through a small grating.

DIFFRACTION GRATING   A plate ruled with 15,000–30,000 equidistant parallel lines used for producing diffraction spectra.

DORÉ   Silver bullion containing gold.

ELECTRON   The smallest atomic particle and the lightest component of matter. It is the fundamental negatively charged element of electricity one or more electrons being present in every atom of all substances.

FERRITE   A magnetic form of substantially pure gamma iron occurring in iron-carbon alloys.

FLUX   Material added to a furnace charge to combine with the gangue in order to produce a fusible slag.

GAMMA IRON   A form of non-magnetic iron, stable between 906° C. and 1403° C. It is the basis of austenite.

GANGUE   Worthless rock and earthy matter associated with valuable minerals in ore.

ION   Electrically charged atoms in solutions or in a gas. Ions are charged either positively (cations) or negatively (anions) according to whether they lose or gain electrons.

KILLED STEEL   Steel that has been deoxidized by addition of manganese, silicon or aluminium before casting to prevent gas evolution in the ingot mould during solidification, sound ingots being obtained.

LIXIVIATION   Leaching; extraction of soluble metals from ore by dissolving in a solvent which does not affect the gangue, the metal being subsequently precipitated from the solution.

MALLEABILITY   The property enabling a material to be mechanically deformed under compression without rupture as in rolling and forging. Gold is the most malleable of all metals.

MARTENSITE   A hard constituent formed in steel by quenching at a sufficiently rapid rate.

MATTE   A mixture of sulphides obtained from the smelting of sulphide ores of copper and nickel.

PASS   A term used to denote the passage of metal between the rolls of a rolling mill in order to reduce the cross-section.

PEARLITE   A microstructure of cast iron and steel composed of a lamellar alternation of ferrite and cementite resulting from the transformation of austenite. So-called from the mother of pearl lustre exhibited by an etched surface.

pH   A symbol used to express the acidity or alkalinity of an aqueous solution on a scale running from one to fourteen, seven being taken as absolute neutrality, values under seven indicating acidity, values above alkalinity.

PRILL   A globule or button of metal frequently found entangled in slags obtained from smelting.

QUENCHING   Rapid cooling of steel by immersion in oil or water in order to harden it.

RABBLING   The operation of raking ore in a roasting furnace, the purpose of which is to stir and move the ore continually, thereby exposing as large a surface as possible to the action of the heat. The mechanically operated iron rakes used in the operation are known as rabbles.

RED SHORT   Brittleness in steels when red hot, usually due to an excessive sulphur content. Also known as hot short.

RIMMED STEEL   Steel that has not been completely deoxidized before casting, gas being evolved during solidification. Characterized by a rim of solid pure steel, the core being of a less pure character and containing blowholes.

SLAG   Non-metallic material formed by the reaction between the flux and gangue of the ore during smelting operations. A refining slag contains the oxidized impurities.

Speiss   A solution of metallic arsenides and antimonides produced in the smelting of ores of cobalt and lead.

Tempering   Operation of reheating steel after quenching in order to decrease the hardness and to reduce stresses.

Tilt Hammer—Helve; An obsolete form of trip hammer used in forging, now superseded by the steam hammer.

Tuyère   A nozzle through which air is blown into a Bessemer converter, blast furnace or cupola.

NAME INDEX

Adams, B. A., 234
Agricola, G., 43, 254
Ahrents, A., 184
Aitchison, L., 19
Anderson, R. J., 226
Arnold, J. O., 322
Ashcroft, E. A., 171
Aston, J., 79

Bain, E. C., 330
Baker, D., 83
Balbach, E., 193, 281, 289
Ball, C. M., 48
Ballot, J., 30, 35, 37
Barba Alvais Alonzo, 285
Bartlett, F. L., 67
Bayer, C. J., 147, 148
Becquerel, A. J., 229
Bedson, G., 310, 316
Bessel Bros., 29
Bessemer, Sir Henry, 1, 64, 65, 72, 75, 94, 96, 221
Betterton, J. O., 192
Betts, A. G., 196
Billings, C. E., 306
Blake, Eli W., 22
Blake, L. I., 51
Boer, J. H., 242
Borel, G., 313
Boss, 286
Bradford, L., 32
Bragg, W. L., 327
Bramah, J., 312
Brandt, G., 255
Bray, C. W., 311
Brearley, H., 4, 125
Brenner, A., 248
Brosse, H. E., 236
Brown, F. C., 276
Browne, W., 203
Brustlein, 124
Buchanan, G. H., 252
Budd, J. P., 81
Bunsen, R. W., 144
Burr, T., 313
Busch, H., 329
Butters, C., 275
Byers, 79
Byrne, S. H., 316

Callow, J. M., 32
Carpenter, J. H., 53

Champion, W., 158
Chance, T. M., 44
Chatelier, H. L., 325
Christensen, N. C., 35
Comstock, 285
Condie, J., 80
Conklin, 44
Coolidge, W. D. L., 319
Cordner, G. D., 248
Cort, Henry, 4, 14, 77, 309
Cottrell, Dr. F. G., 69
Cowles Bros., 145, 146
Cowper, E. A., 81, 103
Cowper-Cowles, Sherard, 177
Crockett, R. E., 50
Cronstedt, Axel, 200
Crowe, 277

D'Alelio, G. F., 235
Danks, S., 78
Dannatt, C. W., 19
Darby, Abraham, 4, 79
Davenport, E. S., 330
Davies, S. A., 225
Davy, Sir Humphrey, 144
Debeye, 327
de Broglie, 329
Delprat, G. D., 30
Dennis, L. M., 252
Deville, H. St. Claire, 144, 318
DeVooys, 44
Dick, G.A., 314
Dings, 50
Dolbean, C. E., 52
Dorr, J. V. N., Dr., 27, 276
Dwight, A. S., 60

Edeleanu, 237
Elkington, J. B., 10, 141
Elliott, 92
Elmore, A. S., 29
Elmore, F. E., 29, 30
Elsner, L., 270
Evans, 58
Everson, Carrie J., 29

Fairbairn, Sir W., 16
Faraday, Michael, 123
Ferranti, S. Z. de, 115
Flanagan, T. J., 199
Forrest, Dr. R. W., 270
Forrest, Dr. W., 270

336

Forrester, 32
Forward, F. A., Prof., 212
Freudenberg, 67
Fritz, J., 310
Froment, A., 29, 30

Gans, R., 234
Gates, P. W., 22
Gerner, 329
Gilchrist, P., 3, 65, 73, 97
Gilson, E. G., 319
Graeff, Dr., 122
Gregor, W., 240
Griswold, 31, 36
Grondal, Dr., 50
Gruson, H., 25
Gwynn, S., 319

Hadfield, Sir R., 3, 123
Hall, C. M., 10, 145, 146, 150
Hall, F., 125
Hall, J., 4, 77
Hamilton, E. M., 287
Hanburg, John, 14
Hardinge, H., 26
Hasenclever, F., 160
Haynes, E., 264
Haynes, Wm., 29
Heberlein, F., 59, 60
Heroult, Dr. Paul, 10, 111, 145
Herreshoff, J. B., 59
Hoepfner, C., 170, 203
Hohlfeld, 68
Holley, A. L., 99
Holmes, E. L., 234
Honda, K., 264
Hoopes, A., 154
Howe, H. M., 322, 329
Huber, 313
Huff, C. H., 52
Hull, A. W., 327
Hunter, M. A., 242
Huntington, T., 59, 60
Huntsman, Benjamin, 1, 4, 94
Hybinette, N. V., 204

Johnson, H. B., 53
Jones, W., 102

Kalling, Prof., 122
Keeney, R., 240
Keith, Prof. N. S., 195
Kelly, W., 1, 65, 98
Keller, C. H., 34
Kjellin, 115
Klaproth, M. H., 229, 241
Klepetko, 58
Knoll, M., 329
Kroll, Dr. W. J., 119, 243, 245
Kron, S. R., 23

Lane, Dr., 158
Langer, C., 205
Lemberg, E., 234
Letrange, L., 170
Lewis, G., 67
Lloyd, R. L., 60
Lodge, Sir Oliver, 68
Lurie, J. S., 44

MacArthur, J. S., 270
MacDougall, 58
MacIntosh, 32
Manhes, Peire, 65, 138
Martens, A., 15, 323
Martin, Emile, 104
Martin, Pierre, 104
Maurer, E., 125
Medina, Bartolome, 283
Menelaus, W., 78
Mergenthaler, O., 301
Merrill, C. W., 277
Meysey, 92
Miller, Dr. F. B., 280
Mishima, T., 264
Mitchell, F. W., 24
Moebius, 281, 288
Moissan, H., 229, 240
Mond, Dr. L., 205
Moore, 275
Morgan, 220
Morgan, S. W. K., 9
Morscher, L. N., 51
Murray, T., 200
Murray, W., 200
Mushet, R. F., 2, 96, 97, 124, 125, 241

Nahnsen, G., 170
Nasmyth, J., 13, 304, 310
Neilson, J. B., 80
Newman, W. E., 180
Nilson, L. F., 242
Northrup, Dr. E. F., 116
Norton, D., 48

Oliver, E. L., 275
Osann, G., 318
Osmond, F., 15, 323

Pacz, A., 157, 269
Parkes, A., 57, 191
Pattinson, H. L., 194
Peirce, W. H., 66, 140
Peligot, E. M., 229
Percy, Dr. J., 78, 127, 128
Perkins, C. L., 34, 35
Peterson, O., 242
Petherick, 43
Pfann, W. G., 253
Picard, H. F., 30, 35, 37
Pilz, B., 182
Polheim, C., 14

Potter, C. V., 30
Price, A. P., 271
Pryce, 216

Quincke, F., 205

Ramsbottom, J., 310
Richards, A. E., 289
Riley, J., 4, 125, 201
Ritchie, S. J., 200
Roberts-Austen, W. C., 15, 325
Rontgen, W. C., 16, 327
Root, E. K., 306
Rosing, 67
Ruska, E., 329

Sauveur, A., 330
Segournet, J., 315
Seltman, Dr. C., 266
Senderoff, S., 248
Sheridan, 31, 36
Sherrer, 327
Siemens, F., 103
Siemens, Sir W., 64, 65, 75, 103, 111, 221
Simpson, J. W., 270, 271
Smith, E. A., 66, 140
Söderberg, C. W., 151
Sorby, H. C., 15, 16, 321
Sorel, M., 176
Strauss, B., 125
Sturgiss, 301
Sulman, H. L., 30, 35, 37

Talbot, H. L., 261
Taylor, F. W., 125
Taylor, J., 23
Tennant, S., 291, 293
Thomas, S. G., 3, 65, 73, 97, 98
Thompson, H. S., 234

Thompson, R. M., 200
Thomson, G. P., 329
Tregoning, H. T., 24
Tromp, 44
Twain, Mark, 286

Uehling, 84

Van Arkel, A. E., 242
Von Bolton, W., 119, 245
Von Laue, M., 327
Von Rittinger, B., 41, 42
Von Widmannstätten, A., 321

Walter, R., 176
Wark, I. W., 36
Watt, J., 4, 80
Way, J. T., 234
Wellman, S. T., 105
Welsbach, Dr., 318
Wenstrom, 50
Wetherill, J. P., 48, 50
White, M., 125
Whitehill, F. T., 35
Whitworth, Sir J., 14, 305
Wilfley, A. R., 42
Wilkinson, J., 80
Wilkinson, W., 299
Williams, J., 275
Williams, J. R., 25
Wilm, A., 155, 156
Winkler, C., 251
Wohler, A., 16
Wohlwill, Dr. E., 281
Wollaston, W. H., 291, 293, 318
Worner, H. W., 248
Wright, Dr., 178

Young, W. W., 72

Leaching
  Cobalt, 262
  Gold, 269ff
  Heap, 12
  Nickel, 212
  Pressure, 212, 262
  Uranium, 232
  Usage, 11
  Zinc, 173
Lead, 177
  Ore-hearth, 178
  Ore-roasting, 185
  Refining, 190
  Smelting, 181
  World production, 198
Light metals, 127
Liquation, 168
Lost wax process, 302

Magnesite refractories, 72
Magnetic separation, 46
  Ball & Norton, 48
  Dings-Crockett, 50
  Grondal, 50
  Wenstrom, 48
  Wetherill, 48
Martensite, 322
Matte, 130, 334
McKinley Law, 221
Mechanical shaping, 297
Melingriffith tinning unit, 224
Merrill Crowe process, 277
Metallography, 15, 321
Metals
  Casting, 13
  Shaping, 12, 297
  Smelting, 62
Mills, Rolling, 14, 306
Mineral dressing, 20
Mixer, hot metal, 102
Moebius electrolytic process, 288
Mond nickel process, 205

Nickel, 199
  Electrolytic, 203
  Mond, 205
  Smelting, 208
Nitrogen in steel, 120

Open hearth furnace, 64
Open hearth steel process, 106
Ore, 20
Ore dressing, 7, 20
Ore hearth, 178
Oxygen, use of in steel making, 120
  Kaldo process, 122
  Linz-Dinawitz process, 121
  Rotor (Oberhausen) process, 122

Pachuca tank, 276
Parkes lead refining process, 191
Parting, gold and silver, 279

Pattinson lead refining process, 194
Patio process, 283
Pearlite, 322
Pig beds, 83
Pig iron, 75, 79
Pig and ore process, 104
Pig and scrap process, 104
Platinum metals, 290
  Occurrence, 291
  Recovery, 292
  Usage, 294
Polling, copper, 132
Powder metallurgy, 317
Pressure leaching, 212, 262
Puddling iron, 76
Pyrometallurgy, 8, 55

Radiography, 328
Radium, 229
Rake classifier, 27
Raschette lead furnace, 182
Recuperation, heat, 2
Refractories, 70
Retorts, zinc, 163
Reverberatory, development of, 133
Reverberatory furnace, 63
Roasting
  Blast, 59
  Fluidized bed, 62
  Heap, 55
  Hearth, 57
  Stall, 56
Roasting furnaces
  Brown, 160, 185
  Bruckner, 186
  Evans-Klepetko, 58
  Hegeler, 161
  Herreshoff, 59
  MacDougall, 58
  Parkes, 57
  Ropp, 160, 186
  Wedge, 59
Rocker, gold, 267
Rolling of metals, 14, 306
Rotor, steelmaking process, 122

Sand and slime gold process, 274
Sand casting, 298
Saniter process, 102
Silica Refractories, 71
Silver, 282
  Arrastra process, 283
  Cazo process, 285
  Chloridizing process, 286
  Cyanidation process, 287
  Electrolysis, 288
  Pan process, 285
  Patio process, 283
  Washoe process, 285
Sink and Float, 44
Sintering, 59

Slag, 63
Sluice, 41
Smalt, 256
Smelting, 62
Smoke farming, 67
Soderberg anode, 151
Solvent extraction, 237
Spitzkasten, 274
Spitzlutten, 274
Stamp mill, 268
Steel, 92
    Alloys, 3, 123
    Extrusion, 315
    Forging, 304
    Rolling, 306
Steelmaking
    Bessemer, 94, 99
    Electric, 110
    Huntsman (crucible), 93
    Open hearth, 103, 106
    Oxygen, 120
Stoves, 81
Strakes, 41
Swagging, 319

Tables, concentrating, 41
Test hearth, 194
Three layer aluminium process, 154
Tin, 213
    Electrolysis, 219
    Refining, 218
    Smelting, 215
    World output, 214
Tinplate, 219
Tinning
    Electro, 223
    Hot dip, 224
Titanium, 240
    Alloys, 248
    Electrolysis, 247

Titanium, Extraction, 242
    Occurrence, 241
    Usage, 249
Tool steel, 3, 124
Transistor, 250
Tops and bottoms Nickel process, 201
Tuyère, 86
Tuyère Injection, 91

Uranium, 229
    Extraction, 231
    Ion Exchange, 234
    Occurrence, 230
    Output, 230
    Solvent extraction, 237

Vacuum
    Dezincing, 192
    Filter, 275
    Melting, 117
Vertical zinc retort, 165
V.L.N. steel, 120

Welsh Copper Smelting process, 129
Wetherill magnetic separator, 48
Wire drawing, 315

Xanthate, 34
X-ray Crystallography, 327

Zinc, 158
    Electrolysis, 170
    Ore roasting, 160
    Refining, 167
    Smelting
        blast furnace, 166
        horizontal retorting, 164
        vertical retorting, 165
    World output, 159
Zone refining, 253